The Metallurgic Age

The Metallurgic Age

The Victorian Flowering of Invention and Industrial Science

QUENTIN R. SKRABEC, JR.

McFarland & Company, Inc., Publishers
Jefferson, North Carolina, and London

Library of Congress Cataloguing-in-Publication Data

Skrabec, Quentin R.
   The metallurgic age : the victorian flowering of invention and
industrial science / Quentin R. Skrabec, Jr.
      p.   cm.
   Includes bibliographical references and index.

   ISBN 0-7864-2326-9 (softcover : 50# alkaline paper)

   1. Technology — Great Britain — History — 19th century.
   2. Technological innovations — Great Britain — History — 19th
century.   I. Title.
   T15.S54   2006
   609.41'09034 — dc22   2005032977

British Library cataloguing data are available

Cover images: *top and bottom* © 2005 Pictures Now;
middle © 2005 Clipart.com

Manufactured in the United States of America

McFarland & Company, Inc., Publishers
  Box 611, Jefferson, North Carolina 28640
  www.mcfarlandpub.com

To Our Lady of Mount Carmel
and my father, who inspired my manufacturing career.

Table of Contents

Introduction   1

1. The Metallurgic Age   7
2. Victorian Incubators—The Great Exhibitions   25
3. Scientific Romances and the Materials Revolution   38
4. Victorian Iron — The Foundation of the Age   51
5. William Kelly — The Inventor of the Bessemer Process   63
6. Victorian Wars and Technology   84
7. Trains and Boats   103
8. Aluminum — Victorian Gold   120
9. Basic Victorian Electricity   130
10. Victorian Metallurgy   147
11. Victorian Toolmaking   161
12. The Victorian Concept of Management   179
13. The Copper Legend — Thomas Edison   193
14. War of the Currents   208
15. The Automobile   223

Epilogue   229
Chapter Notes   241
Bibliography   244
Index   247

Introduction

The Metallurgic Age was the golden age of metallurgy, metallurgical engineering, and industry. Even more, it represented a burst of engineering creativity during the Victorian period (broadly 1800 to 1900) of the industrial nations. It not only advanced metallurgy and metallurgical engineering, but also advanced the science of management that evolved out of the metallurgical industry. "The Victorian period," as used in this book, is a global term for the 19th century culturally rooted in England but later adopted by America as the "Gilded Age." The Victorian period saw the discovery of most metals, their uses, and the prediction of future uses. The use of metals has defined the level of civilization over thousands of years, with such terms as the Bronze Age, Iron Age, and Steel Age. During the hundred years of the Victorian period, the use of metals advanced at an accelerated rate, and so did civilization. Industries developed and organizational infrastructure evolved. Metals made giant bridges, large ships, parts formation, telecommunications, tooling, and mechanical inventions like the automobile.

The advance of industry and metallurgy was directly related to the unequaled creativity of the Victorians. The 20th century advanced metallurgy at a snail's pace compared with the Victorian era. This book looks deep into the connection between Victorian creativity and the advance of Metallurgy, a truly Victorian discipline requiring the application of scientific principles.

The question of why this period caused such a burst of metallurgical advances motivated my research. Many have studied metallurgy from a technical perspective, but few have looked at it from a cultural perspective. The idea of Metal as icon of culture and civilization is unique to the Victorian period.

The Victorian period represents an unparalleled burst of creativity, even from the perspective of the 21st century. Historians have defined great periods of human technology as ages. The period of Victorian creativity is best cast as the Metallurgic Age. Some might argue it was an age of electricity, communication, or transportation, but these are merely the results of Victorian metallurgy. Whereas chemistry dominated the 18th century and physics the 20th century, metallurgy was the heart and soul of the 19th century.

The very first year of the 19th century heralded a new concept: that the coupling of two dissimilar metals could create an electric current. This science of the electrical battery, or "Galvanic cell," would generate the great communication inventions of the telegraph, telephone and radio. The simple Galvanic cell was the basis of electrochemical refining, which led to the discovery of more than 40 new metallic elements before 1840. The Galvanic cell allowed for the process of electroplating as well.

Iron would structure the transportation network that arose with the communication revolution. The iron alloy of steel dominated the physical world of the Victorian period. Steel changed the battlefields as well as the kitchens of the time. After steel, the greatest contribution to metallurgy was aluminum, which was the ultimate product of electro-refining. The incandescent light bulb that changed the world depended on metals such as copper and tungsten.

The accelerated development of metallurgical science was related to several uniquely Victorian attributes: streams of creative thought, amateur scientists, a fascination with engineering triumphs, the evolution of science fiction as a genre, and hero worship. The love of heroes peppered the history books with names such as Edison, Morse, Bell, Bessemer, and Westinghouse, in some cases it seemed the knowledge was searching for the inventor rather than the other way around. Over and over, meta-inventions resulted in patent battles as inventors arrived at the patent office with the "same" idea: Kelly and Bessemer, Edison and Swan, Hall and Heroult, and Gray and Bell. Many times, there were multiple claims, as with the invention of the telegraph, electromagnet, automobile, and others.

These battles extended to the natural sciences. Charles Darwin, for example, rushed his publication of *On the Origin of Species* when Alfred Russell Wallace appeared ready to publish a similar theory. Inventors and scientists, many times totally independent of each other, came to the same conclusions from the evolving stream of knowledge.

This book looks at the contributions and roles of "lost Victorians" who supported the overall advance of metallurgy. Viewing the inventions of the age as streams of knowledge helps one see the heroic inventors of the time in a new light and provides a better understanding of the nature of invention. Victorian scientists added to the body of knowledge, and Victorian engineers applied it just as quickly. The creative environment of the period is not only fascinating but also offers a creative model for any timeframe.

The Victorian period was rooted in enthusiasm and a childlike wonder for science. The new writing genre, science fiction, and international industrial fairs inspired this wonder. Science journals flourished by reporting the new inventions of the day. Many, such as *Scientific American,* got their start merely by reporting new patents. Bigger and better advances in technology captured headlines. In one case, the battle between two behemoth iron blast furnaces captured the headlines in Pittsburgh for months, like celebrity gossip does. The stories of metallurgists and engineers filled the popular press as well. Clubs and associations allowed scientists and amateurs to bond in their love of technology. Openings of new bridges, tunnels, railroads, and new technology became huge public affairs. The popularity of technology and metallurgy led to a blending of industry and the arts. Industry rivaled art as an expression of human creativity. A wave of new fields evolved in the Metallurgic Age, including metallurgy, electrical engineering, automotive engineering, aerodynamics, industrial arts, mechanical engineering, and management science. Technological and metallurgical centers emerged, such as Birmingham, Essen, Pittsburgh, and Manchester.

The period gave rise to a special type of inventor. Rather than coin a new term for this type of inventor, this book uses the term "meta-inventor," adapted from Charles Murray's *Human Accomplishment.* (Murray did not apply the term to physical inventions but to significant ideas. The meta-inventor was the icon of the Metallurgic Age.) The meta-inventor was not a specialist but an enthusiast, able to tap into the streams of knowledge available. Many were amateurs, hobbyists, and experimenters. These lesser-known contributors are truly the "lost Victorians," whose critical contributions are overshadowed by a cultural need to assign a winner. I wanted to tell their stories as well because they were central to the advance of Victorian metallurgy. Another group of "lost" contributors to the Metallurgic Age was the science fiction writers who inspired the use of metals through their works. Jules Verne, in particular, should be credited with the advance of aluminum, steel, electroplating, and metallic batteries.

In engineering, a meta-invention is both a physical and cognitive development, such as Volta's battery. In this engineering definition, we look at meta-inventions as the physical representations of significant and cognitive engineering concepts. Meta-inventions, as used in this discussion, are transforming inventions that lead to advances in technology and civilization. In order to qualify, Murray said these invention must be "followed by transforming changes in practice and achievement." They are not single breakthrough ideas, but deductions from a stream of knowledge. Meta-inventions are more the result of a confluence of ideas and science than an individual idea. Take away an Edison, Morse, or Bell, and the meta-invention is merely delayed, not totally lost. They are also, as Murray defines them, ideas (or, in this case, inventions) that allow humans to "do a new class of things." Hans Christian Oersted's invention of an electrical coil that deflected compass needles, for example, led to the telegraph, telephone, electromagnet, and power generation. A meta-invention, as demonstrated by Oersted's coil, is not necessarily a well-known, popular and historical event.

Another characteristic of the meta-invention is that it occurs at the confluence of streams of knowledge or creativity. The analogy of streams relates to the flow of information in Victorian society. Some may find it difficult to suggest improved flow of information in the Victorian period versus today's Internet age. Today's information, however, is more of a flood than the focused and directed streams of Victorian information. Science fiction writers, scientific societies, the press, and world fairs helped direct creativity. They set goals, digested science, summarized research, filtered information, and directed the streams of knowledge. Inventors such as Hall and Heroult were moving toward commercial aluminum production in a directed manner because of science fiction writers and the press. The search for incandescent lighting was followed and documented over a twenty-year period by the press as well as popular journals. Edison was able to start late and achieve the knowledge level of the competition by searching in books, journals, patents, exhibitions, and the popular press. Such resources helped the dreamers and inventors combine the information for new applications.

The cognitive nature of meta-inventions of the Victorian period can be found in their use by science fiction writers to predict the future. Science fiction writers, for example, had defined the use and applications of aluminum before it was commercially available. The reduction of lightweight aluminum from its ore is a meta-invention that found applications in aircraft, building, and transportation. Electrical current from Volta's battery helped science fiction writers envision a plethora of applications. Similarly, AC power generation was historical node for massive invention. Faraday's discovery of electro-refining opened the new world of metallurgy with the discovery of new metals and alloys. Bessemer's pneumatic steel making changed bridge, ship and building construction forever. Again, the science fiction writers cognitively designed a world of steel long before commercial amounts were available for use. Thanks to the popularity of science fiction, the Victorian mind was racing ahead of physical limitations of science.

Certainly, there were some purely cognitive advances as well. The quest of science led to applications in all forms of human endeavor. Scientific management was at the forefront in these advances. The techniques of management science were critical in the Metallurgic Age. The systemic science of metallurgy lead to the systems approach in other areas of Victorian society. The Victorians were unique in applying the basic principles of science to management. Again, we see the diversity of meta-inventors such as Fredrick Taylor (who discovered high-speed steel alloys) using the same scientific principles to develop scientific management. In fact, the early fathers of scientific management were all from the metals industry. We also see in the development of production techniques another meta-inventor, Eli Whitney. Whitney invented the cotton gin, but his idea of standardization in production is often forgotten. Here again we see the universality of science in the Victorian mind. Other great business leaders such as Charles Schwab, founder of United States Steel, were amateur chemists and metallurgists. As president of United States Steel and later Bethlehem Steel, Schwab pioneered the use of industrial chemistry and metallurgy.

Scientific principles of metallurgy found applications in all facets of Victorian culture. Science bonded a number of Victorian writers outside the genre of science fiction, including Ralph Waldo Emerson, Henry David Thoreau, Edgar Allan Poe, Rudyard Kipling, Charles Dickens, and Mark Twain. Thoreau, in particular, contributed to the development of the manufacture of quality pencils. Emerson, was an effective supporter for the advance of amateurs in scientific matters and engineering. Emerson understood how enthusiasm helped drive creativity in Victorian society. Even Dickens, known for his social satires of the time, was an early believer in and promoter of the use of aluminum in engineering applications. Poe was another Victorian writer who found fascination in science. Poe's love of science is reflected in the work of his biggest promoter — Jules Verne. Science played a dominant role in Victorian society, but the interest had moved from the pure science of the previous century to applied science.

Nations were quick to understand that their status would be based on how well they applied science, metallurgy, and technology. German-born Prince Albert (Victoria's husband) took Britain to the lead role with his promotion of the Great Exhibition of 1851. Britain's dominance in 1851 inspired France, the United States, and Germany to join a technology race. Albert, in particular, focused on the motivation of metallurgical engineers. The race for new metals and engineering applications of them became an international passion. Napoleon III poured money into the development of aluminum and metallurgy to advance France's status, and his awards helped inspire the revolution in steel making. The Civil War further drove the creativity and technological boom in the United States. Abraham Lincoln took the torch from Albert and Napoleon III and stirred a nation into technology advances. Lincoln pushed conservative generals, producing an array of new ordnance, and by doing so led the United States to leadership in tool making and scientific management. Lincoln was the catalyst for the development of the metalworking and machining industry in America. Lincoln, like Prince Albert and Napoleon III, realized that the key to victory on the battlefield was an advance in metallurgy. By the end of the Victorian period, the torch had passed to Bismarck and the Krupp Empire. Metallurgy would define victory on the battlefield as it had in centuries past. Like the Philistines, samurai sword makers, and Toledo sword makers of old, these Victorian leaders made their metallurgical processes national secrets because of their impact on national security.

The history of the Metallurgic Age is a fascinating story, but there is much more. The Metallurgic Age blended science, nationalism, society, culture, and industry into a motivational force to create. For the historian, the study of the period uncovers many forgotten inventors, metallurgists, and scientists. There are corporate models for invention incubators, and employee motivational models that today's manager will find of interest. For the politicians, there are insights into competitive factors for nations in a world market. Even the artist can find much insight and inspiration in the study of this Victorian creativity boom.

1

The Metallurgic Age

The Metallurgic Age is better known historically as the Victorian Age in England and the Gilded Age in America. The Metallurgic Age, however, was not a well-defined time period or world-synchronized time frame. This period of materials engineering flourished in England from about 1830 to 1880, while in America it found its focus from 1865 to 1900. The Victorian and Gilded Age historians view the period from the perspective of dramatic cultural and social changes. The Metallurgic Age arose from the dramatic growth of Victorian chemistry, electrochemistry, metallurgy, and geology. It was a time of science, but more important, it was the pinnacle of applied science and engineering of the scientific discoveries of the previous century. It integrated development of industrial science that produced major advances in metallurgy, geology, industrial chemistry, architecture, structural engineering, and materials science. Metals came to symbolize the age; Victorians used metal for the bridges, machines, tools, buildings, furniture, art, and monuments of their time. Wrought iron and cast iron approached that of gold and silver in art and decoration. Wrought-iron gates and fences surrounded their homes and mansions. Cast iron adorned their building fronts and columns. Cast iron replaced the bronze Napoleonic cannons of the start of the period. Landmarks of the period, such as the Eiffel Tower, Brooklyn Bridge, and the Crystal Palace were monuments of iron. The metals of the future, such as alloy steel, aluminum, tungsten, chromium, and cobalt were freed of their ores, processed, and experimented with.

The period has also been called the Railway Age or Transportation Age since the railroads were the dominant technology of the times. The first train to run was in England and could only do 4 miles per hour. By 1850, a steam locomotive known as the Rocket achieved a speed of 25 miles per hour. The steam locomotive became the start of an almost perpetual economic engine. The steam engine drove up the demand for coal, which increased the demand for rail transportation, which in turn increased the demand for steel rails. Steel rails created new industrial empires of steel to produce them. In 1875, more than 95 percent of all steel production went into railroad rails. The growth of the railroads spurred the production of coal, iron ore, and steel, which then created an economic boom in smaller supporting industries. The very weight of the iron ore that needed to be shipped to the steel mills required the use of steel railroad bridges in place of wood and iron. From 1867 to 1900 in the United States, rail track went from 39,000 miles to 259,000 miles. Steel production in the United States for the same period went from 1.2 pounds per person to 300 pounds per person.[1]

From 1875 to 1900s, the great metallurgic steel plant of Andrew Carnegie in Braddock, Pennsylvania, produced enough railroad rails to lay tracks to the moon and back several times.[2] Growth in the rail system created innovation in all facets of ferrous metallurgy, mining, engine development, and chemistry. Ferrous metallurgy, however, more than other technologies, experienced a boom. Rail technology started the 1800s as cast-iron base technology, moved to better load-carrying wrought iron, improved toughness with the invention of Bessemer steel rails and finally moved to the high quality of open-hearth steel. As with the space program of the 1960s, a lot of the technology was spun off into other areas. Railroad development created steel development, which in turn stirred creativity throughout industry. The search for stronger, tougher, and more wear-resistant steels opened the field of alloy steels using chromium, tungsten, manganese, nickel, molybdenum and cobalt.

The advance of steel-making technology led to a revolution in transatlantic transportation. Steamships greatly reduced the time for an Atlantic crossing. The time went from 23 days eastward and 40 days westward on a sailing ship in 1830 to 5 days either way using steam in 1889.[3] More prized than speed to the Victorians was size. In 1858, England launched the iron transatlantic steamer—Great Eastern. It was known as the "Crystal Palace of the Seas." Historian James Dugan portrayed it this way: "A hundred years ago, in the Iron Age, the world watched a wonder grow at the Isle of Dogs on the Thames in London. It was a steamship five times the size of the biggest vessel afloat then, in the morning of steamers. The colossal hulk was the Great Eastern, mother of ocean liners, the first work of marine architecture to surpass Noah's Ark and the Rhine timber rafts of the Dark Ages. The ship was designed to carry four thousand passengers, almost twice as many as her great grandchild, the RMS Queen Mary, launched seventy-seven years later."[4] Such grand Victorian projects many times advanced technology even though the final product saw limited application. The Great Eastern, for example, was never a commercial success as a passenger liner. Like the supersonic Concord today, the Great Eastern represented the future. Its size would play a key role when the ship successfully laid the Atlantic telegraph cable in 1865. The maiden voyages attracted such visionaries as Jules Verne and Mark Twain and inspired novels and new dreams.

The Metallurgic Age inspired an exponential explosion of metaphysical knowledge and invention. This Victorian period stands as the golden age of innovation, invention, and technology. Based on Charles Murray's recent statistical analysis of human accomplishments, this 19th-century period accounted for 45 percent of significant advances in technology and 53 percent of the significant advances in chemistry from 800 B.C. to A.D. 1950.[5] Murray also looked at "meta-inventors" over the last 2,750 years. A "meta-inventor," such as Thomas Edison or Michael Faraday, is distinguished by a large and significant number of inventions in his field or other fields of endeavor. Of the top 20 meta-inventors, 11 out of 20 are from the Victorian period. The growth of patents in the United States was explosive. In a 40-year period between 1860 and 1900, the U.S. patent office issued 640,000 patents. This was an eighteen fold increase over all the patents issued (36,000) in the 70 years prior. Fully 70 percent of the known metals and metalloids were also isolated during the Metallurgic Age.[6] That is an amazing statistic since the Metallurgic Age accounts for less than 4 percent of the 800 B.C. to 1950 time period. The Metallurgic Age transformed art, architecture, and society as well as industry. Much of its momentum came from the elevation of science to that of a religion. Science was heralded in great exhibitions or world's fairs highlighting science and engineering triumphs. National pride and might was measured by the engineering accomplishments of the nation.

Great scientific and engineering exhibitions, the forerunners of the world's fairs, reached a level of national competition seen today in the World Cup for soccer. They were the technical Olympics of the age. Nations and inventors competed for gold medals. These events raised interest in science and engineering for generations of Victorians and Edwardians. Metals were the center of attention. Metals such as aluminum and magnesium were introduced in the exhibitions of the 1850s. The Great Exhibition of 1851 was a metallic monument of Victorian engineering. The Crystal Palace, the central building of the Great Exhibition of 1851, consisted of 2,300 cast-iron guides and 358 wrought-iron trusses. Cast-iron furniture was found throughout. Some of the most popular exhibitions were a 16,400-pound hunk of zinc, a 24-ton piece of coal, the first metalworking steam hammer, a 32-ton locomotive, the first dinosaur bones, and the world's largest block of cast steel (4,200 pounds). The most visited gold-medal winner was the first steel cast cannon, from Krupp of Germany. The Centennial Exhibition of 1876 ushered in the telephone, electric lighting, and the steel rail. The Eiffel Tower commemorates the Paris International Exhibition of 1889. The Columbian Exhibition of 1893 resulted in the electrification of the world. Over and over, as you trace the history of great inventions, you find the path passing through one of these technical fairs. They were information and knowledge warehouses for inventors, scientists, innovators, and corporations. Many of the world's great companies, such as Westinghouse Electric, United States Steel, Krupp Steel, Libbey Glass, Siemens Electric, and General Electric have their roots in one of these Victorian fairs.

The Royal Academies of Science in England and France further stimulated interest in science and engineering. However, as the amateur scientists and engineers grew, so did Victorian associations and institutions. The British Association for the Advancement of Science (BAAS) of the 1830s was more typical of the Metallurgic Age. While the Royal Academy had grown very aristocratic, the BAAS was more open to the middle class. The BAAS mission was a wider forum for scientific debate and information dissemination. The BAAS met around the country to ensure wider participation. The engineering advances in size, speed and efficacies of machines were followed in the newspaper like the World Series is today. In 1895, the Nobel Prize was created to honor the applied sciences. In the United States, engineering societies evolved to play the same role as the scientific societies. The American Society of Mechanical Engineers (ASME), in particular, helped foster engineering through members, such as Thomas Edison, Alexander Holly, Fredrick Taylor, George Westinghouse, George Babcock, Henry Towne, and Francis Pratt. ASME sponsored conferences for engineers to present their theories. This interest in science and technology created a literary genre of the scientific romance novel (the father of science fiction). It also birthed a group of magazines, trade papers, and journals.

"WHAT WILL HE GROW TO?"

Cartoon from *Punch*, June 25, 1881: King Coal and King Steel looking at the future of electricity.

The *Journal of the Franklin Institute* pioneered the written craft of science and engineering in the 1820s. Its editor was Thomas P. Jones, who in 1828 became the head of the U.S. Patent Office. Jones merged the early *American Mechanics Magazine* with the Franklin Institute's journal and the patent office's journal to create the *Journal of the Franklin Institute.* Jones designed the new journal to "give workingmen plain and practical information and ... direct American inventive activity into constructive channels."[7] Jones envisioned a new type of mechanical artisan born of the American working class. He promoted the term "mechanical arts." Jones believed that uniting the practical skills of the mechanic with the principles of science would give rise to an inventive boom. He did more than anyone else to tie these streams of creativity to national advancement. Jones's goal for the *Journal of the Franklin Institute* echoed Emerson's essay "American Civilization": "Another measure of culture is the diffusion of knowledge, overrunning all the old barriers of caste, and, by cheap press, bringing the university to every poor man's door in the newsboy's basket, scraps of science, of thought, of poetry are in the coarsest sheet, so that in every house we hesitate to tear a newspaper until we have looked it through."[8] Jones pushed hard for the addition of mechanical arts at the high school and university levels. He was successful in creating a new generation of practical scientific and engineering journals.

One of the most successful of these journals was *Nature*, created in 1869 by a group of English scientists. This group was started informally in the early 1860s by an organization known as the X Club. The X Club included Thomas Huxley, Joseph Hooker, Norman Lockyer (editor for fifty years), and John Lubbock. *Nature* proclaimed three objectives: to inform the reading public of new scientific and technological advances, to disseminate new findings, and to allow ventilation of controversies. Its scope was international and tied together the work of science associations around the world. The journal was extremely readable, yet it was a cross between today's *Popular Science* and *Scientific American*. Professor Walter Gratzer at the King's College summarized its impact: "Since 1869, the science publication *Nature* has showcased, exalted, even guided the progress of science, servicing as a forum for groundbreaking research, social commentary, and the occasional offbeat anecdote. H. G. Wells once remarked that no English-speaking scientist could go a week without reading it." The success of *Nature* inspired other international journals such as *Science* and *Scientific American* in the United States. *Scientific American* played an even more active role in science and invention starting in the 1850s. *Scientific American* founded the U. S. Patent Agency to provide technical and legal help for inventors. The journal claims more than 100,000 inventions were patented by 1900 with the help of *Scientific American*. Both Charles Goodyear and Thomas Edison attributed much of their love of science to their early reading of *Scientific American*. Another important function of the journals was that they focused and guided information toward scientific goals. Many of the editors were scientists and visionaries such as Joseph Henry at the Smithsonian. More important, these journals were nests of creativity feeding the next steps of inventions and inventors.

These journals, while scientific and engineering-based, were not controlled by the university-educated but were true journals of creativity and invention. One example of this was the *Monthly Engine Reporter*, which was published every month from 1811 to 1889. Initially the *Monthly Engine Reporter* focused on the performance and application of steam engines in the Cornish mining district. By the 1830s the magazine was covering the broad field of engine performance, new developments, and applications. The magazine adopted the role of testing and comparing engines for industrial applications. This function inspired competition and development among the Victorian inventors. The *Monthly Engine Reporter* played the role of *Popular Science* in the 1960s and 1970s inspiring hobbyists, inventors, and early adopters. There can be no doubt that the Victorian creativity explosion is in part ascribable to the broad involvement of amateurs, tinkers, hobbyists, and students. These types of technical magazines became popular throughout the world, which helped the rapid advance of technology globally.

The Victorian model for innovation and invention represents far more than a piece of period history. The personal computer revolution had an analogous path. Institutional and university science had given the world the transistor and silicon chips, but these scientific achievements had found little new applications. It was a dedicated group of amateurs, tinkers, hobbyists, and scientists whose interaction led to a revolution. Amateur science magazines energized the stream of creativity within this group. Bill Gates attributes the PC revolution to the original articles of hobbyists in *Popular Science*. Small regional computer clubs such as the "Homebrew Computer Club" in California offered an exchange of information. As technology evolved, these groups found more and more applications. New applications brought in professionals from new disciplines, such as accounting. Commercial regional computer fairs came onto the scene. Finally, a group of commercial innovators including Bill Gates, Steve Allen and Steve Jobs emerged to develop the technology further. Another Victorian characteristic of the PC revolution was the aggressive use of technical information and invention. The famous quote of the PC revolution, "Good artists copy and great artists steal," is clearly Victorian. The PC innovators, like their Victorian counterparts, pushed for commercial advantage rather than wait for legal battles. These innovators also lacked university training or even a highly institutional environment. Finally, the technology became integrated in the mainstream.

The Metallurgic Age was an age of globalization. It was globalization that brought together the resources and economic factors to make steel available. This globalization differed from that of today in that cheap labor was brought to the factories, whereas today factories go to the cheap sources of labor. This globalization created massive waves of immigration to America and England. Ireland alone sent millions of immigrants to both America and England. It birthed the mega-cities of London and New York as well as a litany of smaller urban centers such as Birmingham, Pittsburgh and Essen. Industrialization was a source of national pride, and countries preferred to have their industrial assets at home. Britain could have just as easily taken its factories to its huge colonies in Africa, India and Australia, thus avoiding many of the social and environmental problems. But the upheavals in society caused by this quest of industry and science were considered secondary.

The social upheavals were, however, massive. These smoky industrial cities were proud examples of human accomplishment as well as ghettos. Yet, the industrial owners and leaders had emerged from those very ghettos. There was a type of upward mobility that, while limited, is unknown today. Many industrialists, such as Andrew Carnegie and Charles Schwab, had started as uneducated laborers. Great inventors, such as Edison, came from the lower class. The populations of Britain and America tripled during the period. In England in 1801, a fifth of the population lived in cities; by 1900, three-quarters of the population was urban dwellers.[9] Social problems did, of course, follow the rapid industrialization of the period. This more crowded and industrialized society caused historian Clarice Swisher to note, "A sense of faster and more crowded living had its intellectual as well as its mechanical basis. The spread of education coupled with the enormous expansion of knowledge and the corresponding increase of publication, books and periodicals and newspapers, gave 'every man ... a hundred means of rational occupation and amusement which were closed to his grandfather,' and led George Eliot, in a threnody on the death of leisure to say that 'even idleness is eager now, — eager for amusement; prone to excursion-trains, art-museums, periodical literature, exciting novels; prone even to scientific theorizing, and cursory peeps through microscopes.'"

The Victorian period is usually studied and critiqued from the viewpoint of social factors and changes, but this viewpoint misses the heart of the period. For example, many argue that the defining writers of the times were Dickens and Thackeray, but they were merely reporters. It was Jules Verne and Arthur Conan Doyle, with their scientific romances who defined the period. These scientific romances embodied the dreams and vision of this new scientific society, and coupled with the great exhibitions, produced the Victorian incubator for engineering triumphs. They were the DNA of Victorian boys' dreams, which became the engineering triumphs of the next decade. Still, if one looks at a list of significant historical advances of the 1800s, many are noted as having two or more independent inventors.[10] The role of scientific romances that elevated the inventor and scientist to the status of today's sports heroes and rock stars is often understated. Science and engineering can move forward in two different ways. There can be additions of new details to a body of knowledge. This is the methodical approach of today's university-driven science. It is a series of building steps expanding our knowledge or "adding one more brick to the wall." The other path is to open up new areas of investigation based on previous knowledge. This type of Victorian path requires maximum creativity. It rewards boldness and contrary views. It is not absent today, but it is discouraged. This Victorian path cherishes creativity over academic success. It is the heroic science of the Victorian writers, not the directed advance of modern science. It is an approach to science that promotes arguments, feuds and fights and characterized the novels of Jules Verne. It encourages the scientist to break paradigms and alter the direction of scientific research flows. It still depends on the advance of knowledge, but it is not limited by earlier conclusions. Finally, the Victorian creative streams encouraged amateurs and hobbyists to participate in this creative process. This type of popular involvement, with the notable exceptions of the personal computer and astronomy, is missing today in many fields.

It wasn't the university class or old money that drove the industrial revolution, but the rising knights of the lower class. Most of the great inventors and industrialists of the period lacked formal science education. Many of the brightest of the Victorian industrialists realized that the secret of the age was economic motivation. The Horatio Alger model was embedded in the psyches of these Victorians. Many of the period histories stressed rags-to-riches themes. The dream of millions had inspired the persistence of Charles Hall in his aluminum experiments, Carnegie and Bessemer in steel, Morse with the telegraph to name but a few. Andrew Carnegie feared that future generations of his family would lack such motivation. To that end Carnegie, one of the wealthiest men ever to have lived, left only enough money for his wife and children to live their life in style, passing nothing on to future generations. Carnegie thought to do so would rob those generations of the opportunity for success.

In 1878 the *New York Evening Telegram* hailed the patent office as the new American symbol: "It illustrates very strikingly the difference between the fields in which American genius has achieved its greatest triumphs in this century and in the last one. The chief benefits which America conferred upon mankind in the eighteenth century were political. The statesmen of our revolutionary period were instructors of the world in the art of government.... Our national fame in the history of the nineteenth century will rest upon practical discoveries and inventions in natural science and the arts tending to promote the conveniences of life. The Capitol symbolizes American triumphs in the last century, the Patent Office in this."[11]

History has overlooked this churning incubator responsible for an overwhelming number of inventions. Amazingly, many inventions appeared at the same time in different countries. The historical image of the lone Victorian scientist and inventor struggling in an isolated laboratory is incorrect. The Victorian Age consisted of many streams of innovations leading to rivers of invention. The naming of the river usually went to the "map-maker," not the "discoverer." Historians in many cases have forgotten the originators, instead favoring the commercial winners. Inventions such as continuous steel making, plastics, the telegraph, the telephone and the typewriter are credited to the winner of the patent battle and ultimately the economic battle — like Sir Henry Bessemer in the case of steel making — rather than the true inventor — William Kelly. The same is true with the telephone (Bell vs. Gray), plastics (Hyatt vs. Parkes), aircraft (Wright brothers vs. Whitehead), the carbon filament (Edison vs. Swan), radio (Marconi vs. Tesla), the telegraph (Morse vs. Henry), and auto making (Ford vs. Benz), to name only a few. Tesla, Swan, and Kelly all won the final patent battles, but the history books don't reflect the true inventors. These famous patent and commercial battles represented only the visible portion of a much bigger rivalry of lesser-known contributors. In 1886, the breakthrough in the electrolytic processing of aluminum occurred almost simultaneously by Hall in America and French engineer Heroult. The real issue was that Victorian streams of ingenuity could not be stopped. The fact is, without Edison, Bessemer, Morse, and Hall, their famous "inventions" would have hardly been delayed. The advance of basic science was no different. Science advanced more on the competitive battles of scientists than the battles of theories and concepts. Inventive streams of thought would compete to become the headwaters of great scientific and engineering rivers. These great battles filled the scientific journals as well as the trade press. However, our history books remember only the "winners." History has missed the real lessons of innovative streams by recognizing and lifting up individuals. These creative engineering streams were the real touchstone of the Metallurgic age.

In a British Commission review of the textile industry in 1851, these creative streams were identified: "'The present spinning machinery which we use is supposed to be a compound of about eight hundred inventions. The present carding machinery is a compound of about sixty patents.' If we turn to the inner history of mechanical invention, the secret of the great progress is easily learned. As has been shown, there were workers in every field of mechanics, each stimulating and aiding the others, and handing some new facts on to successors. Paul, Wyatt, Kay, Hargreaves, Arkwright, Crompton, Cartwright, and others dealt with the textile problem; Papin, Savery, Newcomen, Beighton, Watt, and perhaps a host of others, whose names will never appear on the pages of history, added to the perfecting of the steam engine." When we fail to recognize these creative steams, we fail to understand the creative process.

These creative and innovative streams flowed because of the advances in theoretical science and communications. In particular, the great science world's fairs brought new concepts and ideas to the streams. There was also travel by industrialists throughout the world to see new innovations. In addition, the detailed writings and reality-based science fiction writers spread the technology. Jules Verne, in particular, used graphic illustration to develop mental pictures of applied science. Verne averaged 70 illustrations per novel![12]

The visionary Scottish mechanical engineer James Nasmyth, who combined his engineering and graphics skills to introduce the field of mechanical drawing in the 1840s, influenced Verne. Nasmyth invented the steam hammer for forging. His steam hammer was on display at the Great Exhibition of 1851, complete with illustrations and drawings. Nasmyth won a prize for that invention as well as a prize for his moon drawings (he was an amateur astronomer). Nasmyth succeeded as an efficient pollinator of Victorian creativity via his travel and drawings. Engineering historian and writer Henry Petroski put it this way: "It is out of such a growing realization of the value of the arts and crafts to civilization and culture that modern engineering started to rise with determination in the eighteenth century. And as the Industrial Revolution gained momentum and technology began to change at a much faster pace than any encyclopedia, whether illustrated or not, could keep up with, there was still no substitute for travel. And it was not only artists who came back from their travels with sketchbooks full of memories and images; the emerging civil and mechanical engineer epitomized by James Nasmyth came back with drawings for new machines and structures."[13]

Because history has overlooked these connections and links in the stories of innovation, many of these innovative Victorians have been lost to history. More important, we are in danger of losing the historical connections, the creative motivators, processes, and networking involved in these advancements. Recent statistical analysis of human accomplishment shows not only the high level of technological achievement during the Metallurgic Age, but also the steady decline since.[14] The Victorian creativity explosion offers both contrasts and insights for today.

One overlooked historical theme of the age was a love and heroic worship of science and technology. In France, the movement had a name — Saint-Simonianism. Claude Henri de Rouvroy, the Count of Saint-Simon, and a socialist who believed the French Second Empire's philosophy was to be transformed by science, highlighted this French movement. The French, as did the British of the time, glorified engineers, scientists and industrialists. Many of the lesser-known writers of the Metallurgic Age believed the real struggle of the period was between technology and man, arts and science, and the environment and industry. Many of the little-known works of H. G. Wells, H. P. Lovecraft and Jules Verne dealt directly with this plutonic struggle of the times. These writers had the pessimistic view that Saint-Simonianism would lead to a final victory of technology over the arts and humanities. The hero worship of engineers had one side benefit in that the engineer ruled in design. When Victorian engineer Isambard Kingdom Brunel built the first great iron cruise liner the *Great Eastern,* he double-hulled it over the objections of investors, at great expense. The *Great Eastern* made its maiden voyage in 1859 with a young writer — Jules Verne. Verne was so impressed by the double-hull safety design he used it as a basis for his fictional submarine, the *Nautilus.* Had early-twentieth-century designers used this approach, tragedies such as the *Titanic* could have been avoided.

It was more than just hero worship of science and engineering; it was a worship of scientific and engineering heroes. Victorian historian Walter Houghton traces the very term hero worship to Victorian times. "Through it has always existed and is still alive today — too much so under Western eyes — hero worship is a nineteenth-century phenomenon," Houghton says. "At no other time would it have been called 'the basis of all possible good, religious or social, for mankind.' In no other age were men so often told to take 'the great ones of the earth' as models for imitation, or provided with so many books with titles like *Heroes and Hero-Worship, Lectures on Great Men, A Book of Golden Deeds, The Red Book of Heroes.* Heroic myth was popular as heroic biography."[15] To the Victorians, scientists, engineers and industrialists replaced the knights and saints of earlier hero worship. The drawback was a desire to appoint a winning name to an invention or theory when it was really a stream of creativity.

Hero worship was of course inspirational, but it hides the advance of science and engineering. Writer Jared Diamond gave a powerful example of this: "In reality, even for the most famous and apparently decisive modern inventions, neglected precursors lurked behind the bald claim 'X invented Y.' For instance, we are regularly told, 'James Watt invented the steam engine in 1769,' supposedly inspired by watching steam rise from a teakettle's spout. Unfortunately for this splendid fiction, Watt actually got the idea for his particular steam engine while repairing a model of Thomas Newcomen's steam engine, which Newcomen had invented 57 years earlier and of which over a hundred had been manufactured in England by the time of Watt's repair work. Nowcomen's engine, in turn, followed the steam engine that the Englishman Thomas Savery patented in 1698, which followed the steam engine that the Frenchman Denis Papin designed (but did not build) around 1680, which in turn had precursors in the ideas of Dutch scientist Christian Huygen and others."[16] The story is repeated over and over as you study the Victorian inventions of the Metallurgic Age.

The National Geographic Society related the following story: "Some years ago, a very high government official sent a draft statement to the Smithsonian Institution for expert review. It included the sentence: 'Every schoolboy knows Samuel F. B. Morse as the inventor of the electromagnetic telegraph.' The experts knew it wasn't so simple as that. They started composing accurate substitutes. Morse invented one of the first telegraphs? A successful telegraph? True, but not dramatic. Finally, triumphantly, they wrote: 'Every schoolboy knows Samuel F. B. Morse as inventor of the Morse telegraph.' Both the official and his audience were happy, and the Smithsonian men could take special satisfaction in the episode because it was their Institution's first Secretary, Joseph Henry, who had laid the scientific basis for a critical part of Morse's work."[17] There is rarely a single inventor, but more commonly one can identify a stream of creativity leading to a single invention. Ralph Waldo Emerson of the period noted a similar assimilation of "new" ideas in his essay "Quotation and Originality":

> The nobler the truth or sentiment, the less imports the question of authorship. It never troubles the simple seeker from whom he derived such or such a sentiment. Whoever expresses to usa just thought makes ridiculous the pains of the critic who should tell him where such a word had been said before.... In fact, it is as difficult to appropriate the thoughts of others as to invent.... It is inevitable that you are indebted to the past. You are fed and formed by it. The old forest is decomposed for the composition of a new forest.

Clearly inventions are mostly the assimilation of older ideas and concepts applied in a new way. The Victorian propensity was to assign a winner or hero to an invention. Charles Murray, in his book *Human Accomplishment*, described the process: "The scientist is engaged in an intellectual Easter egg hunt. Somebody is bound to find any given egg sooner or later.... The rule is harsh rule that serves as a powerful engine for scientific progress: the winner is the one who grabs the egg first, not the one who sees it first. Almost all have heard of Alexander Bell. Almost no one has heard of Elisha Gray. Bell and Gray independently invented similar devices ... but Gray submitted his application for patent two hours later." The explosive inventiveness of the Victorian period supports the idea of intellectual bricklayers coupled with innovative inventors—that is, many inventors contributing bricks to build the overall wall, as well as a wall designer to direct the pattern of bricks. The so-called "meta—inventor, lone genius, giant, mega-inventor, breakthrough inventor, or "system builder" of the Victorian era is indeed rare today, if such a species exists at all. They appear to be more products of the heroic view of history. The term mega-inventor seems to fit best because these great Victorian inventors were diverse professional inventors. Men such as Nasmyth, Bessemer, Babbage, Edison, Faraday, and Morse have a legacy bigger than a single invention. Many of these men had more than one hundred patents in a variety of fields. They were true inventive geniuses, not specialists. In many cases they gained advantage of the specialists in a specific field. They were capable of applying scientific principles quickly to new applications. They were more innovator than inventor, building the work of scientific bricklayers together to create engineering monuments. And they are uniquely Victorian, having no counterparts today.

Materials engineering was at the center of the Metallurgic Age. The steam engine of the previous century was known by many as the iron heart of the Industrial Revolution. It was iron-making technology that harnessed the energy of the great machines of the era. The steam engine was actually invented by Hero of Alexandria in the first century. It failed to develop because cheap slave labor was available to power machines and cast iron was not available to contain its energy. Many historians wrongly identify Watt's steam engine as a breakthrough event. Historian Kirkpatrick Sale said it best: "That may seem like a lot of weight to load on Watt's simple engine—a restatement in metal, by the way, of a device known to the Greeks two thousand years before."[18]

Commercial steel was the material of the age that added new muscle to the Industrial Revolution. Steel is a stronger and tougher version of cast iron. The troika of cast iron, steel and wrought iron (which are in order of lower carbon contents) made up what many called "carboniferous capitalism." The railroads, shipbuilding industry, steel industry and coal industry were locked in an integrated upward economic spiral. By 1880, upturns in any one of those industries translated into upward trends in the others. More than money supply or government spending, carboniferous economic manipulation was the tool of politicians. Iron was the currency of the Metallurgic Age. It even took on status with engineers, some of whom were buried in cast-iron coffins. The age would also raise aluminum to a similar status.

The secret of commercial aluminum was also unlocked during this age with the new science of electrometallurgy. A small amount of aluminum made its debut at the 1855 Paris Exposition. Emperor Napoleon III saw this expensive display of "silver from clay" and ordered a tableware setting to be made of it. This precious tableware set was used for his most honorable quests, while others got gold and silverware. Napoleon III went on to design aluminum armor for his personal guards. French scientist Sainte-Claire Deville gets credit for its discovery and American Martin Hall credit for cheap processing; the real father of aluminum, however, was a forgotten Victorian, William Frishmuth, who commercialized it. It was Frishmuth who first alloyed aluminum with copper and developed casting methods. His precious aluminum pyramid icon was used to top the Washington Monument in 1884 as part of a special lighting-protection system.

This science of electrometallurgy would also lead to the production of magnesium in 1855. The Great Exhibition of 1851 had a major exposition of minerals and ores (on an equivalent basis with machinery). That focus on mineralogy, chemistry and electrometallurgy led to the discovery of cadmium, zirconium, vanadium, silicon, and rare earth metals titanium, cobalt, beryllium and platinum. The pavilion of mineralogy had the largest collection of metal ores ever assembled. The fossil collections of the newly discovered dinosaurs further inspired an interest in geology and science. The central building of the exhibition, the Crystal Palace, became an engineering monument to iron and steel.

Science of the eighteenth century found unusual applications in the Metallurgic Age. Science, as we have seen, permeated literature and even plays. Several of Jules Verne's works were actually popular theater events, such as *Around the World in Eighty Days*. Science took the headlines on a daily basis. Victorian toys were scientific as well. The real advance of science was not, however, from a professional scientific class but from amateurs. Science was the love of the middle class, popular in the great fairs and in the popular literature of the time. Many middle-class managers had a voracious interest in science but in particular the economic branches of mineralogy, geology, chemistry and metallurgy. Many of the great industrialists of the Metallurgic Age, such as steel baron Charles Schwab, were amateur scientists. But it wasn't just industrialists; it was political leaders as well. The scientific interests of Napoleon, Prince Albert, and Abraham Lincoln revolutionized the technology of the world in 19th century.

The real economic success of the period is directly related to the interest of industrialists in the sciences. The period gave rise to a group of entrepreneurial scientists. Charles Schwab (1862–1939), the first president of United States Steel, turnaround president of Bethlehem Steel and financier of companies such as International Nickel, was one of these entrepreneurial scientists. Schwab described that passion: "In my own house I rigged up a laboratory and studied chemistry in the evenings, determined that there should be nothing in the manufacture of steel that I would not know.

Although I had received no technical education I made myself master of chemistry and of the laboratory, which proved of lasting value."[19] Later in life, as a millionaire industrialist, Schwab founded industrial chemistry departments in manufacturing companies and developed metallurgical degree programs at the University of Pittsburgh and the University of Michigan. It was not the universities, however, that produced the advances of the Metallurgic Age, but the basements, barns, and homemade laboratories. These amateurs supplied an enthusiasm, passion, and love that can rarely be found in professional settings.

The Metallurgic Age was a brief period in modern times in which science, not sports, was the world's pastime. Queen Victoria's husband and promoter for the Great Exhibition of 1851 was also an amateur scientist. Many of these amateur scientists changed our world but are lost to mainstream history. The Metallurgic Age's greatest inventors, such as Edison, were self trained engineers and scientists. Victorian historian Walter Houghton described the age succinctly: "The fact is, the Industrial Revolution owed very little to scientific theory. The great inventors—Watt, Stephenson, Arkwright, Hargreaves—had had little mathematics and less science. Their inventions were almost entirely empirical." Upcoming chapters will write the history of many of these forgotten Victorians. One model of creativity that best fits this period is known as the "translation network theory." This model looks at the interaction of scientists, engineers, practitioners, hobbyists, industrialists, and policy makers. The Metallurgic Age represents one of the few periods in history where these diverse groups came into contact via exhibitions, fairs, science fiction, and journals. Today, engineering and science are highly specialized, offering little interaction outside their separate fields.

The works of chemist Humphrey Davy in electrometallurgy led not only to the discovery of new metals such as magnesium and the commercialization of aluminum, but also to the birth of electroplating. Again, industrialists with an interest in science became the innovators of new processes. These metallurgical processes influenced art, allowed for photography, and produced common tableware such as silver plates. The use of gold plating on ink pen nibs by industrialist Sir Josiah Mason revolutionized the art of writing. The pen business was huge in scope. In 1874, Mason was producing more than 4,608,000 pens a week and employing more than a million workers![20] Ink was extremely corrosive to the fine slits in steel nibs. Gold, which is an inert metal, deposited in a thin layer, resolved the problem. This type of explosive creativity crossing into different fields represents the period.

Many of these extraordinary amateurs are well known for other endeavors but forgotten for their scientific advances. Henry Thoreau was one of these unknown industrialists/amateur scientists. Not finding work after graduating from Harvard in the 1830s, Thoreau went to work at his father's pencil factory. Early American pencil makers suffered from low quality compared with German competitors such as Faber. American pencil graphite was "gritty, greasy and brittle." Thoreau spent months researching the chemistry of the graphite, clay and binder mixtures. He was able to identify the key ingredient of the German pencil graphite as Bavarian clay. Adding this new clay produced a harder and blacker "lead," but the product was still gritty. Thoreau resolved this final problem with the invention of improved graphite-grinding machines. Thoreau's work established the dominance of American pencil makers. Within a few years, his breakthrough lead became the basis for a new industry — electrotypesetting. Later in life, Thoreau's interest in pencil making would lead to vertical scientific advances in the understanding of forestry.

Another group of American amateur chemists and metallurgists, known as the "Boys of Braddock," created the Carnegie steel empire and the infrastructure of 19th-century industrial America. The first two presidents of United States Steel Corporation had started like Thomas Edison, with boyhood home chemistry labs. One of these two, Charles Schwab, would forge two other great metallurgical giants-International Nickel and Bethlehem Steel. Schwab, like Mason in pen making and Thoreau in pencil making, mixed science and capitalism to achieve success.

Inventors gained motivation from government, industry, and the hobbyist. When science advanced to the point where no applications could be envisioned, it waited for a practical reason or future vision of use. If, however, there was a clear vision, such as in the future uses of aluminum, it inspired and pulled science forward to meet the need. One of the real success stories of the Metallurgic Age was advancing technology without war. The Civil War and Franco-Prussian wars did result in major advances, but these were no greater than the technological advances of the amateurs of more peaceful periods.

The Metallurgic Age was a cultural think tank driven by economic opportunity. Men like Bessemer, Edison, Bell, Morse and Schwab were motivated by financial gain coupled with a love of science and engineering. This created a scientific industrialism, which caused serious social upheaval. What is overlooked was the general advance of humanity in many areas including the arts and education. Charles Murray's recent analysis of human accomplishments supports a literary burst in the Metallurgic Age. Many at the time believed that technology would ultimately destroy advances in the arts. Even the great visionaries of the time, such as Jules Verne and H. G. Wells, foresaw the future as a technological dictatorship, insensitive to the liberal arts. What evolved were scientific democracies.

The great industrialists, far from becoming technocrats, became patrons of the arts. Industrialists, engineers, and scientists were far from one-dimensional, but were renaissance men of diversified interests. Verne did more than anymore to capture the diversity in his mechanical heroes such as Captain Nemo. There was a concern later in Verne's life and in the writings of Wells that technology and science would overwhelm the arts. It was a prediction that both of the great scientific visionaries had wrong. The spirit of the age brought competition to all aspects of human endeavor.

The Metallurgic Age stands as a great leap forward for mankind. That enthusiasm, connectivity, commercial spirit, competitive nature, and passion has often been lost by its hero worship. This hero worship tended to raise up individual inventors but hide the underlying streams of creativity that resulted in their commercial success. This heroic theory of invention, while inspirational, clouds the history of invention for future generations. Again, historian Jared Diamond has noted this distortion: "The hero customarily credited with the invention followed previous inventors who had similar aims and had already produced designs, working models, or (as in the case of the Newcomen steam engine) commercially successful models. Edison's famous invention of the incandescent light bulb on the night of October 21, 1879, improved on many other incandescent light bulbs patented by other inventors between 1841 and 1878. Similarly, the Wright brothers' manned powered airplane was preceded by the manned unpowered gliders of Otto Lilienthal and the unmanned powered airplane of Samuel Langley; Samuel Morse's telegraph was preceded by those of Joseph Henry, William Cooke, and Charles Wheatstone."

These creative streams of science and technology spilled over into other areas as well. The science-minded industrialists soon turned to the use of scientific principles in the management of men as well as machines. The quest for efficiency naturally led to the human element. The movement started in the early textile industry of the early 1800s, but it soon moved to machine shops and armories. Great machinists, metallurgists, and engineers, such as Eli Whitney, Alexander Holley, Fredrick Taylor, and Charles Schwab, became the leaders in applying science to human production. The new science even took on the name "scientific management." Victorian engineers who learned these practices could benchmark companies internationally. The selection, training and motivation of employees were guided by these scientific principles. Furthermore, the idea of designed experimentation on industrial processes came into vogue. Fredrick Taylor was given the title of "father of scientific management," but Taylor gave credit to 30 years of prior development throughout the world. Applying science to management would also be a base for the development of new fields, such as the social sciences.

The value of looking at this amazing period is to better understand what the causes of advancement were. It offers mankind a model for intellectual growth, commercial expansion and technological development. The 1960s' space program fueled a similar period of technological growth. The 1950s and 1960s saw a surge in science fiction, as well as a revival of Jules Verne's science fiction in movies, such as *Journey to the Center of the Earth, 20,000 Leagues under the Sea*, and *Master of the World*. The same period saw an explosion in scientific toys, such as chemistry sets, rockets, and microscopes. The 1960 New York World's Fair was one of the last great fairs of the Victorian mold, introducing innovations such as the first commercial computer — the IBM 360. Another result was a surge of engineers and scientists that generated innovation and formed the foundation for the race to the moon.

Many historians, such as James Burke, have noted the connections of technological inventions but believed these connections to be randomly evolved. The connections can then later be identified by historical study. Burke describes his pinball theory: "The process by which discoveries are made is almost always turbulent with chance or accident, and only rarely is a discovery the result of careful, rational planning. Science, like everything else in history, depends on how the wheel of fortune spins."[21] The Victorian and Metallurgic Age clearly breaks with this theory of a Darwinian evolution of creativity. While the direction was not formally planned, the science writers, science associations, and world exhibitions directed it. The spirit of the period was unique. It broke the medieval paradigm of limited access to knowledge and learning, as well as the 20th-century paradigm of institutional research. The period shows a bonding between the arts and science under the broader love of human advancement and accomplishment.

Scientists and engineers were generally referred to as philosophers prior to 1839.[22] After 1839, the word "scientist" was coined as analogous to the word "artist." French pioneer of the steam engine E. M. Bataille likened it to poetry: "All great discoveries carry with them the indelible mark of poetic thought. It is necessary to be a poet to create."[23] A Frederic Harrison essay associates the Victorian movement with great musical performance: "Surely no century in all human history was so much praised to its face for its wonderful achievements, its wealth and its power, its unparalleled ingenuity and its miraculous capacity for making itself comfortable and generally enjoying life. British Associations, and all sorts of associations, economic, scientific, and mechanical, are perpetually executing cantatas in honor of the age of progress, cantatas which (alas!) last much longer than three hours.... The journals perform the part of orchestra, banging big drums and blowing trumpets— penny trumpets, twopenny, threepenny, or sixpenny trumpets— and the speakers before or after dinner, and the gentlemen who read papers in the sections perform the part of chorus, singing in unison —'Ah combine Monseigneur[/]Doit être content de lui-même!'"[24]

2

Victorian Incubators—
The Great Exhibitions

The Great Exhibition of 1851 was part science fair, part museum, part amusement park, and part industrial exposition. It was a tribute to the applied sciences and industrial arts that created an era. It represents the real beginning of the Metallurgic Age as a popular scientific movement. The Great Exhibition was the birth of the new sciences of paleontology, geology and metallurgy. From dinosaurs to huge machines, the Exhibition of 1851 spurred the imaginations of a new generation. The famous industrialist and arms maker, Alfred Krupp, never missed a world's fair after seeing the success of 1851. Within 25 years the world would see the invention of continuous steel making, the telegraph, a massive railroad expansion, and the commercialization of new metals such as aluminum, tungsten, nickel and others. The telephone, electric light, and automobile would appear at the end of this same era. The Victorian inventions would define our modern world. The creativity and innovations of the Victorians would create a new industrial culture. There is no other period of history that would see so many diverse inventions and rate of industrial growth. These world exhibitions created even more competition and achievement from year to year. While similar to the space race of the 1960s and the computer revolution of the 1980s, the Victorian knowledge advanced across a broad spectrum of science, engineering and the arts.

The roots of the British great exhibition go back to the national French Industrial Exposition of 1844, as well as some earlier French and Prussian industrial fairs. The project gained momentum with the energetic interest of Prince Albert, the German husband of Queen Victoria. Albert was an amateur scientist and national visionary whose love of technology projected England to the forefront of technology. The press, however, opposed Albert and the exhibition initially. The term "Crystal Palace" was actually a derogatory press description. Prince Albert had a passion for science and technology that would prevail. He was a daily visitor to the exhibition and attended most of the lectures. Albert weathered the initial critics to be bathed in praise. The exhibition was a popular success as well as a financial success. Profits from the fair went on to build the Albert Museum and Imperial College. The theme was to be "The Great Exhibition of the Works of Industry of All Nations." Eventually 32 nations would participate, making it the first "world's fair." To ensure an international theme, the British Commission opened design and construction bids to all nations. In the end France won more of the construction contracts than England. The exhibition had more than 6 million visitors. The royal visitors reflected the international favor: Napoleon III, Tsar Alexander of Russia, the shah of Persia, the kaiser of Germany, the queen of Greece, the sultan of Turkey, the khedive of Egypt and many lesser dignitaries. Science was the star of the fair. It was an amazing exhibition known for the first full reproductions of dinosaurs and the Rosetta Stone and Koh-i-noor diamond. The public, however, would take more note of a cast-steel cannon in the German Krupp exhibit. Along with the cannon, Krupp had a world-record piece of steel. Krupp's success at the exhibition ensured his company's presence and sponsorship for the next 50 years.

Walter Houghton defined the mind of Victorians at the exhibition:

> [I]t was ... an excited tribute to [the] power of man to conquer nature; to the human mind that could discover her secrets and transform her material resources into productive usefulness. "Within the last half century," wrote one Victorian in the year of the Great Exhibition, "there have been performed upon our island, unquestionably, the most prodigious feats of human industry and skill witnessed in any age of time or in any nation of the earth." Equally significant, because his attitude toward science and industrialism was not uncritical, is Carlyle's excited recognition of the "wonderful accessions," which the century was making to the physical power of mankind. "We remove mountains, and seas our smooth highway; nothing can resist us. We war with rude Nature; and, by our resistless engines, come off always victorious, and loaded with spoils." The same spirit infected the manufacturers themselves. They seemed to Mrs. Gaskell "to defy the old limits of possibility, in a kind of fine intoxication, caused by the recollection of what had been achieved, and what yet should be.... There was much to admire ... in their anticipated triumphs over all inanimate matter at some future time, which none of them should live to see." That fine intoxication was only possible in days when big business and applied science were still young, and the material world was all before them.[1]

It was clearly a time of excitement and enthusiasm for technology.

Another prime mover in the creation of the international industrial fair was the Royal Society for the Encouragement of Arts, Manufactures and Commerce (RSA). The RSA was formed in 1754. The society had encouraged the invention of practical appliances and machines using prize money and awards. The RSA is credited with bringing England to the forefront of the industrial revolution. Prince Albert's biographer, Hermione Hobhouse, summarized its impact: "The prize-winning appliances make a fascinating list. Drill-ploughs, root slicers and chaff cutters were symbols of the Agricultural Revolution which created so much of the prosperity which enabled England to stand alone against Napoleon. Looms, cranes and pumps were of equal use to the growing host of mechanized industries which created the new towns of the Midlands and north-west."[2] The RSA emerged nine years prior to the Royal Academy, but by the 1840s, it had fallen on hard times. Prince Albert became president of the society in 1843, turning the fortunes of the organization. One of the early historians of the society noted, "He impressed on the Society that the main object of its existence henceforth must be the application of science and art to industrial purposes; and the Society soon learned that it could expect from him, not merely a formal consideration of routine matters, but a quick and wise judgment on any important question that was submitted for his opinion."[3]

To meet the expectations of Prince Albert for a massive international event would require a scale never seen before. Albert and the society proposed a mall of buildings and structures. The main building was to be the Crystal Palace, which would cover 772,784 square feet, or about six times the space of St. Paul's Cathedral. The design was by Sir Joseph Paxton, an architectural amateur in conservatories and greenhouses. The huge glass structure contained 3,500 tons of cast iron, 550 tons of wrought iron and 900,000 feet of glass. Besides attacks by the press, the Astronomer Royal, Professor Airey, questioned the design. The concerns centered on the modern disease of resonance — the idea of vibrations caused by crowds resulting in collapse. Several bridges of the period had collapsed because of resonance, and there was a fear that any large structure might suffer this end. Particularly noteworthy was the collapse of the English cast iron-wrought-iron composite bridge. Then in 1850, an iron French bridge collapsed, and 478 soldiers were thrown into the Marne at Angers. Both of these disasters had captured headlines throughout Europe. Tests on the palace structure included 300 workmen jumping simultaneously in the air. Finally, to simulate regular oscillations, an army drill team marched repeatedly across the structure. The success of the experiments eliminated public fears.

The selection of Joseph Paxton as architect was typically Victorian. Paxton was a gardener and amateur scientist. He had no professional credentials in engineering, architecture, or science. Paxton was an established horticulturist, having published many books on the topic. His love of gardening helped him incorporate plants and trees into the design of the Crystal Palace. The Crystal Palace with its skylights, trees, and plants was a precursor to the modern shopping mall. Part of this was to calm objections to cutting down a group of old elms at the Hyde Park location. Its unprecedented design success was attributed to Paxton as the inspired amateur so typical of this age. The design was to be copied throughout the world. Paxton was knighted for his design.

The palace also had other unusual design features. It incorporated prefabricated cast — and wrought-iron sections for the first time. Hawks were used to control the sparrow population. Public flush toilets were used for the first time. The building was sprayed with water for cooling and included mechanically adjustable louver shutters for control of air circulation. Canvas shades were required at times to reduce the heat of reflected light. A six-inch water main around the building with fire cocks supplied water for the event. Steam power was used for machinery. Overall this exemplified an engineer's exhibition showcasing the triumph of engineering methods. England supplied the machinery, France took the gold medals, Germany awed the world, and the Americans showed them all how to apply those advances to manufacturing. The famous Thomas Jefferson–Eli Whitney demonstration was repeated for exhibition visitors. Parts of ten muskets were mixed up and then randomly selected and the muskets reassembled. Technology historian Donald Cardwell noted, "Whatever impressions the general public took from the Great Exhibition, British engineers like Joseph Whitworth, concerned with machine tools and guns, were much impressed by the Colt revolver and the Robbins and Lawrence rifle with interchangeable parts."[4]

Exhibitions included the McCormick Reaper, a submarine, artificial legs, powdered graphite yellow pencils, Lucifer matches and new automatic controls for looms. The concept of the dinosaur was introduced to the public with life-size models. Fossils of the dinosaurs of England proved popular and led to the formation of the National History Museum. Stuffed elephants and other African animals drew the same huge crowds as the dinosaurs. Some unusual items included household furniture made from coal, centrifugal pumps, an early version of photography known as camerae-obscurae, and electric clocks. A unique exhibit of hydrogen-filled rubber balloons by Charles Goodyear took a medal. The enthusiasm of the visitors led to the development of a National Science Museum. Steel and steel making captured the imagination of engineers as well. One of Prince Albert's favorite exhibitions was the mineralogy hall, which had metallic ores from all over the world. It inspired Albert to become a patron of the newly formed Museum of Practical Geology. Albert ultimately combined this museum with the College of Science to form the Government School of Mines. This was the first college to offer courses in metallurgy. Albert also encouraged evening lectures to be open to the public, and Albert often attended these lectures. That enthusiasm continues to be implanted in future scientists visiting these institutions today.

The great Krupp crucible steel-casting "monster" weighing 4,300 pounds, was considered a technological miracle in 1851. It required the simultaneously poured product of ninety-eight crucibles and disciplined working crews and managers. The achievement inspired Jules Verne in his novel *From the Earth to the Moon* (1870), to fictionally cast 114,000-pound cannon using 1,200 furnaces tapped simultaneously. Verne dedicated a chapter to describing the casting—"The Festival of Casting." Alfred Krupp's casting was a great metallurgical achievement. Krupp had visited England early 1851 to see the British deliver their world-record crucible steel casting of 2,400 pounds. Krupp viewed the arrival of the British casting as a challenge to Germany and his steelworks. That challenge inspired the Krupp "miracle" casting. Krupp's casting awed the public with their love of records and bigness, but it would be two other steel entries of Krupp's that would change the world. His steel railroad tires and steel cannon anchored a new world of technology that would alter the very nature of society.

The impact of the steel cannon and Krupp's steel-making technology cannot be overestimated. The German ingot alone represented more than the total production of steel in the United States in 1850. Kelly and Bessemer had not yet fully developed their experimental pneumatic steel. The world's militaries were stuck on Napoleonic bronze cannons because of their reliability and ease of production. Bronze cannons had won at Waterloo, which left an imprint on the world. Krupp's cast-iron cannons had failed during the Thirty Years War, and that had also left a mark. Softer wrought-iron cannons had fared no better publicly. In 1844, a 12-inch smooth-bore naval gun on the USS *Princeton* exploded, killing the American secretaries of State and Navy. The Prussian military even rejected Krupp's early steel cannons of the 1840s. Krupp continued to improve his casting techniques and steel quality prior to 1851. Krupp's steel, however, offered the optimum tradeoff of toughness and strength. At the exhibition, Krupp's six-pound cannon under the Prussian flag saw a stream of military men and politicians, including Queen Victoria and her master general of ordnance. Another almost daily visitor was a young inventor—Henry Bessemer.

The core of the Great Exhibition of 1851 centered on the celebration of man, the toolmaker. Machinery Hall had endless rows of mundane machine tools. But these long rows were crowded with machinists, toolmakers, gun makers, and the world's military officers! The famous machinist and father of the computer, Charles Babbage, wrote in his biography of the exhibition, "It is not [a] bad definition of man to describe him as a toolmaking animal. His earliest contrivances to support uncivilized life were tools of the simplest and rudest construction. His latest achievements in the substitution of machinery, not merely for the skill of the human hand, but for the relief of the human intellect, are founded on the use of tools a still higher order." In 1851 Britain held the title of the world's toolmaker, but it was the Americans who applied those tools to advance manufacturing. Donald Cardwell in his history of technology, *Wheels, Clocks and Rockets*, describes the situation: "The manufacture of muskets and shotguns in Britain was in the hands of the craftsmen of Birmingham, the city of a thousand trades. The individualistic Birmingham craftsmen relied on the manufacture of sporting guns to earn a living. Manual skill, not advanced machine tools, was the practice, even perhaps the boast, of the Birmingham gun trade; it was accordingly quite unable to meet, cheaply and rapidly, a sudden rush of orders from a government facing a national emergency. The Great Exhibition, if nothing else, must have convinced the British army that there were methods of producing better guns more quickly and more cheaply. And the place to find out these methods was the United States." The American methodology not only increased gun-making productivity, but also created the new science of management.

Another debated impact of the exhibition was the rise of functional as well as decorative designs. Eclecticism is the term most often used to describe Victorian design. Eclecticism became the act of borrowing from many sources to satisfy oneself. This borrowing at times caused overdecoration. For example, Rodgers and Sons of Sheffield displayed a knife with eighty blades. In addition, the knife was inlaid with gold with detailed etching. Victorians loved gold painting, plating, inlaying, and gilding. While beautiful, it lacked function and was clearly overdecorated. Still, it celebrated the purely Victorian in that it was high technology and the triumph of manufacturing. Victorian design, while between functional and decorative, did break new ground in the area of instrumentation and machines. Prince Albert did more than any other Victorian to link art and industry. Ralph Waldo Emerson said, "The English love the lever, the screw, and the pulley, and as early as 1759 the economist Adam Smith stated, "Utility is one of the principal sources of beauty." The illustrations in the works of Jules Verne and H. G. Wells show the same overdecoration of instruments and equipment. Jules Verne's *Nautilus* was an amazing example of Victorian eclectic design, as was Wells's *Time Machine.* The overdecoration seems almost fitting because the beauty was in the equipment. Victorian historian Clarice Swisher summarized this way: "British engineering was the best in the world, encouraged by the efforts of men like James Nasmyth, the inventor of the steam hammer, and his pupil Joseph Whitworth, whose machine tools at the Great Exhibition were praised for their great beauty and power. Machine castings used Greek friezes and columns to decorate the tooling. The nineteenth century toolmakers have been acclaimed by modern art history as a new race of artists, and even at the time, objects like lathes and microscopes were considered beautiful as well as useful; more significantly, many appreciated that they were beautiful because they were useful."[5] The term "industrial arts" was solidly embedded in the culture of the Victorians after the Great Exhibition.

Another Victorian historian went on to suggest the Great Exhibition changed the work ethic: "The Great Exhibition, which provided an opportunity for much debate on design, also created a new attitude to work, which was to prove equally significant. Work had become respectable; trade and manufacture were given the nation's blessing. Albert himself, who could have chosen to be little more than an idle and handsome figurehead, was much inspired by the romance of industry, and worked hard himself, contributing to the view that hard work is good even for the richest of us. In the words of *The Economist* of 1851, 'Labour is ceased to be looked down upon ... the Bees are more considered than the Butterflies of society; wealth is valued less as an exemption from toil, than as a call to effort.'"[6] Industry and manufacturing became inspirational, and many Victorians linked it to the Protestant work ethic.

The success of the Great Exhibition, even though France had won the war of medals, still appeared as a setback to Napoleon III because England had hosted this first truly international fair. Napoleon III rushed to the challenge, proposing the Universal Exhibition of 1855. France missed the point of an international fair by controlling the contracts. It was a French affair, but like the Great Exhibition, it did inspire technological competition. While the Great Exhibition predicted the future of steel, the Universal Exhibition of 1855 featured the future of aluminum. In fact, aluminum stole the show. It was a world challenge to free aluminum from its ore. Napoleon III became world promoter of aluminum. Jules Verne incorporated the use and future applications of aluminum in many of his novels, originally inspired by his visit to the Universal Exhibition. The international quest for aluminum reduction can be traced through the world exhibitions of the 19th century. Exhibitions continued, but it was not until the U.S. Centennial Exhibition of 1876 at Philadelphia that the world's fair image was recaptured.

Unlike the previous great European fairs, the Centennial Exhibition of 1876 centered on an existing building—Independence Hall. It was the first large-scale exhibition in the United States and heralded America's emerging industrial muscle. In 1874, President Grant formally invited the world to participate. The exhibition produced fewer breakthrough technologies than the Great Exhibition, but it was a far greater tribute to machinery of the age. Machinery Hall housed the world's largest steam engine, which every day drove 75 miles of belts and shafts. Machinery Hall was typical of machine shops throughout America. Lathes, presses and forges crowded the hall. Machinery Hall is preserved today as part of the Smithsonian Institution. There were also several of machines that harbingered the communications revolution. These were Christopher Sholes's Remington typewriter and Alexander Graham Bell's telephone. Like Alfred Krupp, who overwhelmed industrialists with his steel cannon at the Great Exhibition, steel maker Andrew Carnegie captured their imagination with the steel railroad rail. The Bessemer steel rail of Carnegie Steel Corporation was 120 feet in length and was the longest rolled steel rail ever at the time. Krupp Steel wowed the crowds with the world's largest cannon, a 35.5-centimeter-bored monster capable of firing a 1,000-pound shell. This prodigious gun was given to the Turkish sultan after the exhibition. Krupp also exhibited a number of peaceful applications of steel. Krupp had learned to use these world fairs to gain marketing advantages throughout world's steel markets.

The 1876 Exhibition was considered the birthplace of American engineering. Alexander Holley found inspiration from the exhibition to form several engineering societies. Holly had previously been involved in the formation of a group of Victorian metallurgists in 1871. That group would become the American Institute of Mining, Metallurgical, and Petroleum Engineers. ASME historian Frederick Hutton noted that influence in his 1915 look at the role of the exhibition:

> The Centennial Exhibition in 1876 in Philadelphia was responsible for a quickening in mechanical matters and for a growing sense of latent power. The big central Corliss engine of Machinery Hall was a splendid object lesson and this Exhibition was signalized by [a] single valve automatic engine with flywheel governor designed by John Hoadley[, by] Professor Sweet's design of the straight-line engine, and by series of broiler tests by Charles E. Emery, Charles Potter, and Joseph Belknap. These all marked epochs in the engineering history of the United States. Moreover, in the fifteen years since the Civil War the enormous increase in size and Productivity of industrial plants had just begun.

The Professor Sweet mentioned was John Edson Sweet, an educator and inventor. He had caused a stir at the Paris Exhibition of 1867 with the introduction of a linotype machine. He loved the information exchange so prevalent at these exhibitions and wanted a more routine vehicle for engineering exchanges. Meetings in 1876 between Sweet and Holley would ultimately lead to a new engineering association. Sweet teamed up with Holley to form the American Society of Mechanical Engineers in 1880.

The Centennial Exhibition ushered in the new electrical age, but few visitors probably grasped the beginning of this age. As night came each day, the lamplighter would make his rounds to ignite the gas streetlights. Injured Civil War veterans were often employed as lamplighters, which was the origin of the term the "lame lamplighter." Horses and horse-drawn carriages filled the streets. Kerosene lamps lighted houses. Blacksmiths were still common and necessary to keep people moving. General Electric Company historian John Winthrop Hammond described the dawn of the electrical age:

> Eighteen seventy-six was America's centennial year, and its Centennial Exposition was destined to open a new gateway, on which might well have been carved a proverb after the manner of Solomon: "It is the glory of God to conceal a thing; but the honor of man to search it out." Through that gateway lay the Electrical Era.
>
> The prophets of that age smiled in condescension when its pioneers dared to assert that the future belonged to the mysterious force with which Ben Franklin, more than a century before, had played. People might listen; but soon they shrugged skeptical shoulders and turned to the other wonders of the Centennial Exposition, marking the end of a century of existence for the United States of America.
>
> The Exposition, like all such displays, was chiefly a record of things done. It was concerned with history, not with prophecy. Nonetheless, there was a real prophet present, not of flesh and blood, but a thing of iron and copper; not a man with ringing voice, but a man-made machine, whispering a new language. That prophet was the electric dynamo.

The next grand international fair after Philadelphia was the International Exhibition of Paris in 1889. This fair commemorated the centennial of the French Revolution. It is of course best remembered for the Eiffel Tower, a monument to wrought iron shooting 300 meters (about 985 feet) into the air, at the time the tallest structure in the world. As a statement to the Metallurgic Age, the tower's height was greater than any that could be built out of stone masonry. Its construction was riveted wrought iron containing 2.5 million rivets and 15,000 iron pieces. It required forty tons of paint. It was designed and built by Gustave Eiffel. At the time, wrought iron was well established as a structural building material, but steel had been making inroads. By 1883, the Chicago Home Insurance building was the tallest in the world (10 stories) using wrought iron for the lower levels and Bessemer steel for the higher. Still at the time of the 1889 Parisian exhibition, wrought iron appeared a safer bet for the record-setting tower. Eiffel had wrought-iron bridge experience as well as having just completed the structural framework for the Statue of Liberty.

As with the Crystal Palace, the press opposed Eiffel's tower, not so much for its construction as its lack of beauty. Of course, we know that ultimately Paris would love the design and retain it as a monument. The attacks were smart, however, calling it "an offense to French good taste." Author Alexandre Dumas called it "baroque, mercantile imaginings of a machine builder." Dumas clearly missed the nature of Victorian design, but Eiffel did not in his response to Dumas: "Can one think that because we are engineers, beauty does not preoccupy us or that we do not try to build beautiful, as well as solid and long lasting structures? Aren't the genuine functions of strength always in keeping with unwritten conditions of harmony? ... Besides, there is an attraction, a special charm in the colossal to which ordinary theories of art do not apply."[7] This "special charm" of metals and machinery was the very essence of Victorian design. The love of Eiffel's design continues to this day; who could visit Paris without visiting the Eiffel Tower? Today the top of the tower includes the restored office of Gustave Eiffel and a premier restaurant — the Jules Verne. Besides the Eiffel Tower, the International Exhibition of Paris heralded electric lighting. Lighting was supplied by 1,150 arc lamps and 10,000 incandescent lamps for about a million candlepower of light. The major attraction was a giant incandescent Edison light in the American Pavilion. The hero of the Paris Exhibition was Edison with his endless inventions and ideas. The Siemens Electric Company pioneered the use of electric streetcars.

Edison and the Americans would usher the United States into leadership at the Paris Exhibition of 1889. The *New York Daily Tribune* declared in 1889, "America could have in Europe no worthier representative of the consummate flower of its National life and progress than this modest scientific investigator and industrious mechanic. Its chief contributions to the world's stock of civilization have been the works of its inventors. In that beneficent field of human effort its sons are unrivalled for practical skill, habits of scientific investigation and triumphs of mind over material forces. While the European Continent to-day is a circle of camps swayed by the caprices of sovereigns whose inherited functions are their only title to fame, America has expended its energies in working out an industrial development that is the marvel of Christendom, and the real leaders of its pacific progress have been and are its inventive mechanics— men of the Edison camp."[8]

While the Great Exhibition of 1851 ushered in the future of Victorian accomplishments; the 1893 Columbian Exposition at Chicago celebrated the successful completion of those accomplishments. The central monument of the Columbian Exposition was both mechanical and structural. In addition, it was a steel structure. This superstructure would become known as the Ferris wheel. Even more amazing was that it was a moving superstructure powered by electricity. Today's readers may write this engineering marvel off as an amusement park invention. The engineering behind it is the real story. First, the wheel itself was 250 feet diameter, its nearest competitor in 1892 being the 72-foot water wheel on the Isle of Man. The axle was the largest steel-forged shaft in the world. It had been made at Andrew Carnegie's Homestead Works, which was the largest forge in the world. The wheel consisted of 36 cars, each the size of a bus holding 60 passengers. The design required six cars to be loaded at a time, which required six stops per revolution. Once the cars were loaded it made one full revolution at a cost of fifty cents to the passenger. That was about a twenty-minute experience for the passenger. In all, 1.5 million passengers rode the wheel in six months of operation. It remained the largest Ferris wheel ever built until 1982, when a Japanese wheel claimed the record (the Chicago wheel was destroyed in 1906).

Washington Gale Ferris, a graduate engineer of Rensselaer Polytechnic Institute, designed the wheel. He had gained experience as a bridge engineer and inspector. In the 1880s, Ferris formed the construction company of G. W. G. Ferris & Company in Pittsburgh. This firm was a leader in steel bridge construction throughout the United States. This experience would be critical as the wheel was built in components using various Pittsburgh mills. Testing and inspection was done in Pittsburgh and Detroit, while pre-assembly was done in Detroit, and then sections shipped to Chicago. In one month alone, 150 railroad cars of components were shipped from Detroit to Chicago. The shipping of the massive axle required a special railroad car designed to move the cannons of Krupp in Germany. The Ferris wheel did what all fair projects had done; it stretched, advanced, and celebrated technology.

The Columbian Exhibition is probably best remembered as a great electrical exhibition. Its lighted midway is often remembered even today. The exhibition featured arc lighting, incandescent lighting, fire alarms, telephone service and electric transportation. While the Paris Exhibition had pioneered the use of lighting at a fair, Chicago overpowered the French. The Columbian Exhibition had 90,000 incandescent lights and 5,000 arc lights for a total candlepower 11 times that of Paris. The source of power was a Westinghouse alternating current generator, which had won the battle over the Edison direct current lighting. In addition to lighting, these Westinghouse generators operated an in-park rail transportation system. Edison had won over Paris a few years earlier, but Chicago was to be the triumph of George Westinghouse. Another prominent electrical event was the opening of a New York-to-Chicago telephone wire by Alexander Bell.

The awe and amazement of the electrical applications can be seen in this report of the opening of the fair:

> It is electricity that whirls the chariot wheels — the thunderbolts are harnessed at last. It is the same sorcery that day and night tell the wondrous story by telegraph and telephone to the ends of the earth and will yet signal the stars in their courses, that carries orders and rings alarms; that bids the nations of the earth good evening and good morning. There is a map showing the electrical features at Jackson [P]ark, and the simple recital of the items shows their strange variety, and in how startling a degree they are comprehensive. The

whole electrical service at the exposition comprises these systems: arc lighting, incandescent lighting, electric power, telephone service, police signal service, fire alarm service, telegraph service, electric transportation! ... The adaptability of electricity to the service of man has daily development, and it gains incessantly new territories of usefulness. It was necessary to place lighted buoys for seven miles along the shores of Lake Michigan, from the Chicago [R]iver to the grounds of the exposition, to indicate the shoals, and there are thirteen spar buoys.[9]

Two companies built massive pavilions of their own. One was Libbey Glass, which built a complete glass factory managed by the future inventor of the bottle-making machine, Michael Owens. The other was the seasoned fair exhibitor Alfred Krupp. Krupp featured his new nickel-steel armor and his chrome-nickel-steel cannon. Both these products would change the world in just a few decades. Krupp biographer William Manchester summarized Krupp's exhibit: "Krupp expended a fortune on the Chicago World's Fair of 1893. To house his displays he constructed a separate pavilion, a replica of Villa Hugel [Krupp's Ruhr Valley mansion] with his name emblazoned across the façade. It cost $1,500,000. The *Scientific American* devoted an issue to it. 'Of all the foreign nations that are taking part in the World's Columbian Exposition at Chicago,' the front page article began, 'Germany takes the lead, in extent, variety, cost, and superiority in almost every characteristic. Of the private exhibitions, Krupp, the great metal manufacturer of Germany, stands at the head. His exhibit is wonderful, and by its greatness almost dwarfs all other exhibits in the same line.' Drawings showed 'a cast steel bow frame for a new German ironclad,' 'one of the Krupp traveling cranes, used for slinging and moving the great Krupp guns,' and three 16.24-inch cannon mounted on hydraulic carriages. 'A special interest attaches itself to this particular gun,' one caption read, 'because it was tested in the presence of the German Emperor at Meppen on April 28, 1892. On that occasion it was fired nearly thirteen miles.' "

One of the most underrated exhibits, the exhibit of the newly formed Pittsburgh Reduction Company, turned out to be a harbinger of the future. It used the same Westinghouse generators that the fair used for lighting and that also opened the technology for the commercial production of aluminum. The company displayed the original pellets produced by Charles Hall in a shed at Oberlin College. There were a few examples of other manufactured aluminum products. The leading investor of the Pittsburgh Reduction Company, Captain Alfred Hunt, was, however, disappointed in their exhibit. The following year, Hunt teamed up with two other Pittsburgh industrialists, Andrew Carnegie and George Westinghouse, to hold a regional fair — the Pittsburgh Region Industrial Exposition (1894). This regional exposition allowed for the exhibit of aluminum bicycles, pots, pans, and aluminum's use in steel making. The Pittsburgh Exhibition would launch a movement toward area fairs in the years between the larger international exhibitions. Andrew Carnegie, after the Chicago fair and regional Pittsburgh fair, was convinced that America needed these expositions for the advancement of technology.

The Chicago World's Fair more than any other event showed the dominance of America in the industrial arts. It also heralded an era of American globalization and the rise of manufacturing exports. Fair historian Reid Badger summarized the impact:

> Widespread admiration was expressed for American technology, factory management, inventiveness, organization, mass production, and especially for the extensive and novel uses of electricity. Foreign visitors were also impressed by the short hours and high pay of American workers and by the apparent importance attached to women in industry and the arts. Several European nations (France, Germany, and Great Britain) were represented by notable displays of their products and skills, but the contrast with the American exhibits overall was not as great as at the Centennial seventeen years earlier, and in the mechanical arts and sciences the United States was shown no superiors. The fair appeared in this sense a triumphant celebration of the age of the machine, and it did stimulate American entrance into the international commerce in goods on a more extensive scale than ever before. "It is a well known fact," wrote Ferdinand Peck in 1899, "that more American firms have been able to form connections abroad and extend their foreign trade since 1893 than ever before."[10]

As you read the many reports of how this fair launched a world market for America, you can't help thinking it is time for another great exhibition in America.

3

Scientific Romances and the Materials Revolution

More than anything else, it was the visionaries who spurred the Victorian scientific and engineering revolution. The French author Jules Verne (1828–1905) defined the space program a hundred years before the technology was actually available. Verne started a creative stream of engineering that birthed submarines, aluminum buildings, the telephone, airplanes and the Apollo moon project. It was the great Victorian science fiction writers who are truly the forgotten contributors of the age. They played many roles in the creative streams of ingenuity. These writers in many cases filtered and focused scientific information for inventors. Many of Verne's novels discussed in detail the development of scientific ideas. They also motivated and inspired the learning of science. Verne more than any other writer was part of the explosion of scientific enthusiasm. He was not merely a science fiction writer but the inventor of the genre known as scientific romance. Verne educated, inspired, and planted dreams. He didn't just write about the great scientific, national, and engineering competitions of the age; he actually precipitated them. A hundred years later, Wernher von Braun stated that many of the American space innovations were directly attributable to Verne. Actually the list of those to credit Verne publicly is a long one: Guglielmo Marconi, Admiral Peary, Admiral Byrd, Robert Goddard, Simon Lake, Yuri Gagarin, Neil Armstrong, Frank Boreman, Walt Disney, Arthur Clarke, Cyrus Field, and even Pope Leo XIII. The real power of Verne's influence was in tens of thousands of lesser-known scientists and engineers who were part of the deep Victorian streams of innovation.

Jules Verne is a name that has survived generations, but the name of the man behind it has not. Pierre-Jules Hetzel played the role of publisher, censor, publicist, nationalist, educator and scientific visionary for Verne. Scientific visionary not in the sense of an icon like Verne, but supporting science as a foundation of civilization. Hetzel was depressed by France's performance in the Franco-Prussian War of 1870 as well as by the new government's lack of an educational plan for the nation. Hetzel's fixation on the French state of affairs and his proposed solutions mirrored the sentiments of the public at large. "To provide his nation's young people with the cultural and educational means to reassert their collective identity, regain their pride, and match the technological advances of their conquerors became a top priority, not only to Hetzel but to the entire country."[1] Hetzel was the true Victorian in that he saw science as part of society. Without science, Hetzel believed France would become a second-rate nation. This belief is seen in an 1870 letter to his son:

> Our poor France! Well, the greater her misfortunes, the more we will love Her, and more! It's our generation that has allowed her to fall into the Abyss; it's yours, my child, that will pull her out of it. And military strength is not the first weapon to be used. No. It is first through the combination of education and learning that we can put this defeated country back on track. Have we lacked guns or personnel? No. It's Science ... and, above all Discipline.

Hetzel's hand would weigh heavily on all of Verne's writings. He proposed to Verne the concept of a series of science romances to be known as the Voyages Extraordinaire. The project, Voyages Extraordinaire, would build a generation of new French scientists and engineers. This project would consume Verne's writing life. Hetzel, as Verne's publisher and editor, had from Verne's first works promoted education. Prior to the series, Verne had achieved fame for *20,000 Leagues under the Sea*, *Five Weeks in a Balloon*, and *Journey to the Center of the Earth*; this series would define Verne as a writer. The series would begin with *Mysterious Island* (1874) and end with *Master of the World* (1904). The series remained educational with as many as 20 to 30 pages of a novel dedicated to explaining scientific principles or the history of scientific ideas.

Hetzel, in the 1874 preface to Jules Verne's Voyages Extraordinaire, explained the ideology of the next 30 years:

> The excellent books of M. Jules Verne are part of a very small number of those that one can give to the new generation with absolute confidence. In the contemporary market, there exist none better to answer society's needs for learning about the marvels of the universe. There exist none which have been more justifiably greeted with instant success from their very first appearance.
>
> If the public's fancy sometimes wanders toward works that are flashy and unwholesome, its basic good taste never permanently settles on any work that is not fundamentally wholesome and good. The two-fold merit of the works of M. Jules Verne is that the reading of these charming books has all the flavor of a spicy dish while providing the substance of a nourishing meal.
>
> The most respected critics have acclaimed M. Jules Verne as a writer of exceptional talent who, from his earliest works, has made a place for himself in French letters. An imaginative and exciting storyteller, a pure and original

writer with sharp wit and, in addition, a profoundly learned author, M. Jules Verne has succeeded in creating a new genre. What is promised so often and what is delivered so rarely, instruction that is entertaining and entertainment that instructs, M. Verne gives both unsparingly in each one of his exciting narratives.

The Novels of M. Jules Verne have moreover arrived at a perfect time. When one sees the general public hastening to scientific lectures given all over France and that, in the newspaper, art and theatre columns are making way for articles on the proceedings of the Academy of Science, one must conclude that Art for Art's Sake is no longer enough for our era. The time has come for Science to take its place in the realm of Literature.

The merit of M. Jules Verne is to have, boldly and masterfully, taken the first steps into this uncharted land and to have had the unique honor of a well-known scientist say of his works: "These novels will not only entertain you like the best of Alexandre Dumas but will also educate you like the books of Francois Arago."

Young or old, rich or poor, learned or uneducated, all will find both pleasure and profit from these excellent books of M. Jules Verne. They are sure to become friends to the entire family and will occupy a front shelf in the home's library.

The illustrated editions of the works of M. Jules Verne that we are offering at an unusually low price and in a luxurious format show the utmost confidence that we have in their value and in the ever-growing popularity that they will achieve.

New works of M. Jules Verne will be added to this series, which we shall always keep up-to-date. All together, they will fulfill the intent of the author when he chose as their sub-title: "Voyages in Known and Unknown Worlds." The goal of the series is, in fact, to outline all the geographical, geological, physical, and astronomical knowledge amassed by modern science and to recount, in an entertaining and picturesque format that is his own, the history of the universe.

Hetzel clearly saw the series as an educational project. Verne had already demonstrated a passion for scientific writing. Verne honed his edge by constantly reading and staying current on the advances of science. In *20,000 Leagues under the Sea*, for example, Verne gives a basic lecture on electrometallurgy embodied in the story. In *From the Earth to the Moon*, Verne describes the production of cannon in terms of the latest technology as well as predicting the use of aluminum in space. In *Mysterious Island*, Verne's teaching on steel making and glass making is equal to some handbooks on the subject. As Hetzel suggested in his preface, Verne does this in a seamless manner, never lecturing, but integrating it in the story. Hetzel edited harshly, keeping Verne on the educational path. Science was always to be the central "romance," and moral themes of Verne suffered the blue pencil of Hetzel. Hetzel even put certain novels of Verne on hold to intensely focus Verne on science and education. One of these 100-year-old manuscripts, *Paris in the Twentieth Century*, was just recently published!

Part of this educational aspect of Verne's novels was the use of graphics and illustrations. Hetzel and Verne pioneered the pedagogical device of illustration in fiction. Maps were used through the Voyage Extraordinaire books to teach and inspire imagination. Drawing of new designs and engineering detail were common in Verne's works. Overall, Verne used 4,500 illustrations in his Voyage Extraordinaire series, or 70 per book. These illustrations were extremely popular with the Victorian reader. Verne would also pioneer the use of footnotes in fiction. Both of these pedagogies were used to educate the reader in a unique way. Many of the dime novels of the time expanded the use of both. Verne's educational success can be seen in editor John Lockyer's review of Verne's *Off on a Comet* in *Nature* (1877): "These remarkable works, which we owe to the genius of Jules Verne ... are well deserving of notice at the hands, for in the author we have a science teacher of a new kind. He has forsaken the beaten track, *bieu entendu*; but acknowledging in him a traveled Frenchman with a keen eye and vivid imaginations— and no slight knowledge of the elements of science — we do not see how he could have more usefully employed his talents. We are glad to have such books to recommend for boys' and girls' reading. Many young people, we are sure, will be set thirsting for more information."

Another Victorian author of scientific romances was H. G. Wells. Wells managed to create scientific heroes, but his novels were not as educational as Verne's. Still, Wells's characters were just as inspirational to young scientists. He used much less science in his works trying to be more predictive than Verne. That lack of science and engineering grounding made him less prophetic. Wells foresaw the rise of the automobile, but missed the airplane's role in transportation. Wells had predicted the demise of the horseless carriage as a result of its filth and pollution. Verne was a science reporter first, and that allowed him to identify creative streams in science. Wells tended more toward conservative fantasy.

Verne alone was capable of taking basic science and extrapolating to technological breakthroughs. He developed into a serious student of the Metallurgic Age. Verne became in effect the first professor of metallurgy for the age. He probably deserves to be called a pioneer of aluminum metallurgy. In his novel *From the Earth to the Moon* (1865), Verne uses a roundtable design scene to interject a metallurgical lecture. His characters, President Barbicane of the Baltimore Gun Club (an artillery engineer), General Morgan (a ballistics expert), and J. T. Mason (an armor expert) debate the merits of cast iron, copper and aluminum in a lunar projectile. In the end the decision is based correctly on aluminum's strength-to-weight ratio. Verne even predicted the exact height and weight of the Apollo space capsule 102 years before man's landing on the moon. He saw the ultimate path of science, even though at the writing of the novel, aluminum had a cost greater than gold and the amount Verne predicted would be needed far exceeded that in existence in 1865 (literally a handful). Verne played a critical function in the technological advances of the Victorian Age, and with others, focused the science and gave direction to the engineering. Technology was no longer part of an earlier pinball or random series of connections, but a targeted advance. Readers of Verne, such as aluminum reduction inventors Charles Hall and Paul Heroult, were focused from an early age on a goal of commercial production of aluminum.

In 1855, Verne had personally viewed the world's inventory of manufactured aluminum — a few pellets— at the Paris Exhibition. The exhibit was called "Deville's silver" after French scientist and industrialist Sainte-Claire Deville. Deville had improved on the older German process, but neither process could produce more than a few pellets. Still, the idea of a process that would produce "silver from clay" caused a sensation. Another visitor at the Paris Exposition would further advance the production of aluminum. That visitor was Emperor Napoleon III of France. Napoleon III became fascinated with the precious metal. Initially he had aluminum buttons produced for his coat. As more aluminum became available, he had a helmet made of it. His coat was the envy of other European rulers, and Napoleon III, like his great uncle, loved to show off. He then ordered a table setting of aluminum spoons and forks for his better quests. Lesser guests were forced to eat with gold and silver tableware.

Napoleon III proposed to go even further, suggesting equipping his army in aluminum armor. Such a project's costs at the time could be compared to those of President Kennedy's proposed trip to the moon in 1962. Napoleon III had to first supply capital to build the first commercial aluminum plant. The plant never reached the emperor's vision because of cost overruns. The plant was built at La Glacierre, but because of pollution the local residents forced it to be moved closer to Paris. Verne had the same enthusiasm as Napoleon III for the future of aluminum. In this case, however, it was the scientist Deville who was the source of predictions for the writers. Deville's writing in the 1850s not only detailed his experiments and processes, but also discussed the future uses of aluminum.

Yet another exhibition visitor and Victorian writer, Nikolai Chernyshevsky, was infatuated with the metal. Chernyshevsky was a lesser-known Russian writer of scientific romances. The following is from his novel *What Is to Be Done?* (1863): "How light is the architecture of this inner house, how small the piers between the windows, and the windows themselves are huge, wide and tall, reaching to the ceiling.... But what are these floors and ceilings made of? And the frames of these doors and windows? Silver? Platinium? Oh, I know now, Sasha showed me a plank made from this material and it was as light as glass. Now they wear such ear-rings and brooches, yes, Sasha said that sooner or later aluminum will replace wood and maybe even stone. How rich it all looks! Aluminum everywhere ... Here in this hall half of the floor is bare, and you can see that it is also made from aluminum." Chernyshevsky writes here two years before Verne, showing that innovative streams existed even in literature.

Verne, however, did not limit his use of aluminum to space exploration. In *Master of the World* (1904), Verne had Robur's airplane incorporate it into the plane's design. H. G. Wells also predicted the use of aluminum in aircraft in his short story "A Dream of Armageddon." Verne foresaw the evolution of aluminum in aircraft. Verne's earlier novel *Clipper of the Clouds* has his character Robur build their first plane out of wood and cloth. Later novels would propose aluminum as replacing wood. Again, Verne's vision proved accurate with airplane skins being made of aluminum by the 1930s. Even the Wright Brothers (fans of Jules Verne's) used it in the first airplane with an aluminum engine crankcase to save weight. Verne goes even further in *Propeller Island*, prognosticating the use of aluminum in buildings and skyscrapers. Verne not only saw the future of aluminum, but he became a supporter of steel as well.

Verne's early work *From the Earth to the Moon* (1865) contained many pages of discussion on the use of cast iron, wrought iron and steel. In *Mysterious Island*, Verne created a basic textbook for ferrous metallurgy. Verne was early to claim the praises of steel. He correctly foresaw the superiority of forged steel over wrought iron in armor. He discussed in detail the accomplishments of American ordnance engineers— Parrott, Dahlgren and Rodman. He compared them to their European counterparts— Armstrong, Palliser, and Tresisilles de Beaulieu. For political reasons, he chose to ignore the Krupps of Germany. Both Verne and Hetzel deplored the militarism of the Prussians. This was such a glaring omission that in 1874 a translator, Edward Roth, added the Krupps to the text discussion. Verne dealt with the politics of the great arms races of the world in his novel *The Begum's Fortune* (1879). In this negative framework, Verne addressed the technological success of the Krupp cannons, which had burnt France badly in the Franco-Prussian War.

In the novel *From the Earth to the Moon*, Verne chose the hollow casting process developed by General Thomas Rodman to produce his giant cannon for the moon shot. Verne discussed the advantages of this new process. Rodman had produced the Civil War's largest cannon — the Columbriad. The Columbriad was a cast-steel cannon produced in Pittsburgh in 1861. It had a 20-inch bore and was hailed by the United States as the largest in the world. Verne's research showed the largest to have been a Russian gun with a 36-inch bore cast in the 16th century.

Steel was a favorite material of Verne's, and Hetzel saw the importance of its development to France. In *20,000 Leagues under the Sea* (1870), Verne predicted the use of steel in ships. The submarine of the novel, the *Nautilus*, was formed of steel. Verne prophesied the further use of steel in shipbuilding in his novels *Floating City* (1876) and *Propeller Island* (1896). Steel was Verne's vision, but he found inspiration in the use of metal in shipbuilding in the Victorian behemoth the *Great Eastern*. Verne's steel *Nautilus* submarine is an example of how he assimilated engineering advances into his fictional creations.

Consider the following description of the *Nautilus* from *20,000 Leagues under the Sea:*

> The Nautilus is composed of two hulls, one inside, the other outside, joined by T-shaped irons, which render it very strong. Indeed, owning to this cellular arrangement it resists like a block, as if it were solid. Its sides cannot yield; it coheres spontaneously, and not by the closeness of its rivets; and the homogeneity of its construction, due to the perfect union of the materials, enables it to defy the roughest seas.
>
> These two hulls are composed of steel plates, whose density is 7.8 that of water. The first is not less than two inches and half thick, and weighs 394 tons. The second envelope, the keel, twenty inches high and ten thick, weighs alone sixty-two tons.

Verne, a voracious reader of technical journals, seems to have used the concepts of two American engineers. The first was Henry Shreve, who designed twin-hull wooden steamboats for the difficult navigation of the Mississippi River in the 1830s. The second was James Eads (father of the steel bridge and ironclad gunboats), who designed "submarine" iron diving bells using double hauls in the 1850s. More recent research suggests Verne had a real live model for his *Nautilus* as well as his main character — Captain Nemo. He was also well aware of the double hull of Brunel's *Great Eastern*.

That lost piece of history is a fascinating story of a forgotten Civil War submarine and President Lincoln. The story begins in Verne's hometown of Nantes, France, around 1835. An eccentric engineer, M. Brutus de Villeroi, had designed a wrought-iron submarine. Verne would have been seven years old in 1835, but being a daily visitor to Nantes's docks, he would have known of the submarine. The cylindrical submarine measured 33 feet long and 4 feet wide, making local history in France. Villeroi had submerged with a small crew for three hours. The ship had some very unique features, such as an air lock to allow crew members to leave the ship under water. The ship was also propelled by a screw propeller. (American Robert Fulton had taught the use of a screw propeller submarine to the French. Fulton had been contacted by Napoleon in 1801 to build a prototype submarine, that he named the *Nautilus*. Fulton demonstrated its success, but it was never adopted.) The most unusual claim of Villeroi was a secret chemical process for the creation of an artificial atmosphere. The feat seems to have been "lost" to most until 1861.

In 1861 President Lincoln received a long letter from Villeroi. Villeroi detailed the following possibilities for his submarine's use: "[to] reconnaissance the enemy's coast, to land, ammunition, etc., at any given point, to enter harbors, to keep up intelligence, and to carry explosive bombs under the very keels of vessels and that without being seen. With a few such boats, maneuvered each one by about a dozen men, the most formidable fleet can be annihilated in a short time."[2] This appeared to be the very weapon Lincoln needed in a stalled-out war. Lincoln personally sent the letter to the Navy Department with a handwritten endorsement.

The Navy officers went to see the vessel in operation, which was later called the *Alligator*. Villeroi won a contract to supply the Union an adapted version. It is not clear how much of the technology, such as the special breathing apparatus and the screw propeller, actually existed, but the contract called for both. It's uncertain whether Villeroi actually had the special breathing equipment. Some claim that a dispute with a subcontractor prevented the breathing equipment from being installed. In any the case, the *Alligator* was delivered without the breathing equipment. Villeroi also failed to deliver on the screw propeller. The *Alligator* had a "duck foot" system of oars in place of a screw system. Without the breathing system, the *Alligator* was extremely limited if not crippled. Villeroi disappeared from France, and the Navy Department had a turkey. The *Alligator* was reworked by the Navy Department and made some voyages in the James River during 1862. Eventually, during a heavy gale the *Alligator* was lost in tow off Cape Hatteras in April 1863.

The Confederates, however, can point to *Hunley*, which didn't really fare much better. The *Hunley* used a screw propeller and was able to achieve five miles per hour. The Hunley was a wrought-iron tube based on an early but forgotten iron sub known as the *Plongeur Marin*. The *Plongeur Marin*, built in 1850, used a screw system. The *Hunley* used an eight-man crew with no breathing apparatus. Even without breathing apparatus, the eight-man crew was able to stay submerged for more than two hours. A candle was used to monitor the oxygen level while submerged. This performance was close to that of the *Alligator* and further suggests Villeroi had no secret equipment. The *Hunley* emerged as an offensive ship versus the more defensive role Villeroi proposed for the *Alligator*. The *Hunley*, using a pole with a torpedo attached, was able to sink a Union ship. The Union *Housatonic*, one of the Federals' most powerful ships, was blockading Charleston in 1864, and the *Hunley* successfully sank it. The *Housatonic* was able to drag the *Hunley* down too. The Confederates built another submarine, the *David*, but it saw little action. The *Hunley* had created fear in the North. The idea of a ramming-type submarine like Jules Verne's *Nautilus* was behind that fear. The federal Navy had started late in the war a hand-propelled screw sub known as the *Intelligent Whale*. Lack of interest following the end of the war ended its development, but this experimental craft can be seen at the Navy Shipyard in Washington.

Verne in his 1870 novel, *20,000 Leagues under the Sea*, resolved a number of the *Alligator*'s technical problems, at least fictionally. In the novel Captain Nemo explains his breathing apparatus this way: "I could manufacture the air necessary for my consumption, but it is useless, because I go up to the surface of the water when I please. However, if electricity does not furnish me with air to breathe, it works at least the powerful pumps that are stored in spacious reservoirs, and which enable me to prolong at need, and as long as I will, my stay in the depths of the sea." We now know that the inventor of the 1850 French sub, Wilhelm Bauer, created a sub in 1863 that used compressed-air bottles. This may well explain Villeroi's secret as well. Verne also chose to use screw propulsion for his submarine, as Nemo again explains: "The electricity produced passes forward, where it works, by electromagnets of great size, on a system of levers and cog-wheels that transmit the movement to the axle of the screw. This one, the diameter of which is nineteen feet, and the thread twenty-three feet, performs about a hundred and twenty revolutions in a second." Villeroi's submarine, of course, did not use electricity to drive the screw but manpower. This concept of electrical power Verne anticipated from seeing the first electric dynamos at the Paris Exhibition of 1867. Verne added steel to the design and developed a long-range submarine for underwater travel. He used the Victorian engineering of the *Great Eastern* as well.

The *Great Eastern* was one of the great Victorian engineering spectacles. Conceived in the year of the "Crystal Exhibition" (1851) and launched in 1857, it was followed actively by the press of Europe. One of its first passengers was a young Jules Verne, taking a rare international trip. The size of the ship earned it the names "Floating City" and "Crystal Palace of the Seas" in the press. Verne would title two of his novels *Floating Island* and *Propeller Island*. Press analogies of size were typical in Victorian speculation. The size was compared to the combined tonnage of 197 British ships that fought the Spanish Armada.[3] Another often-used comparison was to Noah's Ark. Sir Isaac Newton, a century earlier, had calculated the ark to be 515 feet long and 86 feet wide with a tonnage of 18,231. The *Great Eastern* was 693 feet long and 120 feet wide with a tonnage of 22,500. The *Great Eastern* was twice as large as any existing ship of the time. Its double hull represented the triumph of engineering over cost. In addition, the ship had both a screw propeller and a side paddle. Isambard Kingdom Brunel built it out of riveted wrought iron. Verne, while copying the double-hull design, improved on it, using steel plate.

Brunel was the archetypal Victorian engineer who had gained fame for his cast-iron dome design of the Crystal Palace. The *Economist* of today described him this way: "Isambard Kingdom Brunel, pictured famously with a cigar in his mouth and mud on his trousers, probably spent more time talking to parliamentary committees than designing railways. Many of the great civil engineering projects of the nineteenth century were run by engineers who also had to be designers, managers and entrepreneurs all at the same time. Modern engineering consultants are technicians as different from nineteenth-century entrepreneurs as a cameraman is from a film-producer."[4] Brunel, like his ship, offered Jules Verne another model for his fiction.

The real motivating power of these scientific romances was not in their predictive accuracy, but in the creation of scientific heroes. Hetzel's demands and editorial control for educational purposes maybe missed the real impact of writers like Jules Verne on the youth. Verne and Wells created illustrious professors, scientists, and engineers. They were far from the dull principles they mastered. These fictional characters were colorful adventurers and masters of the world. They traveled the world beyond as Phileas Fogg in *Around the World in Eighty Days*, Professor Lidenbrock in *Journey to the Center of the Earth*, and Captain Nemo in *20,000 Leagues under the Sea*. They conquered the world as Verne's Robur in *Master of the World* and Harry Killer in the *Barsac Mission*. Likewise, H. G. Wells created models for Victorian scientists such as the Time Traveler (*The Time Machine*) and Mr. Cavor (*First Men in the Moon*). Verne made it clear that to enter this league of exceptional men first required knowledge of science. Science would bring wealth, status, and fame. The scientific romances of the time created a new type of hero for a society that already cherished hero worship. These Vernian heroes were well educated in the sciences and the arts.

Victorian historian Walter Houghton saw hero worship as a fundamental characteristic of the age. Houghton describes the evolution of hero worship: "When the Victorian period began, all the prerequisites for hero worship were present: enthusiastic temper, the conception of the superior being, the revival of Homeric mythology and medieval ballad, the identification of great art with the grand style, the popularity of Scott and Byron, and the living presence of Napoleonic soldiers and sailors. But traditions die without nourishment, and this one throve. For it answered, or it promised to answer, some of the deepest needs and problems of the age. In fifty years after 1830 the worship of the hero was a major factor in English culture."

Scientific heroes were multidimensional, unlike the military heroes of the age. Literary critic and editor Arthur Evans illustrated this with Verne's Nemo: "Captain Nemo, perhaps the most famous of Verne's heroes, is remembered foremost as the designer, builder, and master of the Nautilus. But his engineering and technical skills, although impressive, are not what truly bring him to life. It is rather his intrepid willingness 'to go where no Man has gone before,' his intriguing personal past shrouded in mystery, his rebelliousness, his implacable sense of social justice, and his passionate love of liberty and the sea. Nemo is the epitome of the 'Renaissance man'— one who excels not only in intellectual and physical pursuits but in cultural ones as well. Note, for example, the titles of those works contained in his 12,000 volume shipboard library. One finds therein not only the great books of science, but also 'all the ancient and modern masterpieces.'" This was the scientist hero of Hetzel's vision — an educational inspiration. With Hetzel's death in 1886, Verne's scientist heroes became more negative in their misuse of science, but still fascinating to the reader. The power of Verne's role models can be seen in the explosion of scientists as a result of the 1950s Verne revival in the movies. By the 1950s, Verne's predictions had lost their awe, but his heroes hadn't.

Verne was not alone in his characterization of the scientific hero. Arthur Conan Doyle developed Sherlock Holmes and Professor Challenger. Holmes, the amateur chemist, typified the Victorian love and application of science in all facets of life. Science was at the heart of many of Holmes's adventures. Professor Challenger was the lead character of Conan Doyle's *The Lost World* (1912). Challenger captured the Victorian battles of science. Challenger launched an expedition to South America to find a lost world of dinosaurs. Again we see the scientist as a heroic adventurer so common in Verne's novels. *The Lost World* was one of the many Victorian variations of Verne's *Journey to the Center of the Earth*. Some, like Sam Moskowitz, a science fiction literary critic, see Professor Challenger as the more powerful characterization: "The main character, George Edward Challenger, if not the finest drawn character in all science fiction, is at least on a par with Verne's Captain Nemo, Burroughs' John Carter, and Stanley G. Weinbaum's alien entities. The dumpy, barrel-chested, black-bearded, bad-tempered, intolerant, egotistical, driving, but truly brilliant Professor Challenger, despite his faults, or possibly because of them, bubbles into believability from the black type of the printed page." Challenger was pure Victorian, a composite of people like Darwin, Huxley, Brunel, Bessemer, Edison and others.

Verne and Wells were of the very top of the scientific literary stream that was inspiring American youth. At a lower level, the dime novel was promoting science and invention. Actually, in the 1880s these novels sold for a nickel. Many times they were exact storyline knockoffs of Verne's works. Some, however, borrowed themes but developed their own characters. The little-known author of these pulp novels was Luis Philip Senarens;

in fact he published most of his life under the title "Noname." Senarens perfected the scientific Horatio Alger story for popular boys' magazines such as *Boys of New York*. Senarens earned the title "the American Jules Verne" while publishing scientific stories for boys.

This amazing man was reviewed in Sam Moskowitz's book, *Explorers of the Infinite*: "If you are wondering how you could conceivably have overlooked so many vital points in American history, the answer is probably that you never read the dime novels which were so popular in the latter half of the nineteenth century. Hundreds of 'untold chapters in history,' similar to the foregoing, appeared in them, seemingly an entire school of literature. Actually, more than 75 per cent of all the hundreds of the prophetic dime novels written during that period were the work of a single man — Luis Philip Senarens — concealed beneath the masquerade of 'Noname'; and he might have remained anonymous but for the premature report of his death issued on the eve of the hundredth anniversary of the birth of Jules Verne in 1928." Senarens did have a relationship with Jules Verne, with whom he exchanged letters in the 1880s. Both writers seemed to have a creative, cooperative relationship, sharing ideas and themes.

Senarens used not only Jules Verne as a model but also Thomas Edison. Later in his career, Senarens developed a series known as Tom Edison, Jr., which found great popularity. Edison enjoyed good press throughout his life and loved to add to his own legend. Edison's life was a popular article in any newsprint. Senarens was able to tap into the public desire to know more about Edison. His main series, however, was closer to Verne's themes. This series was known as Frank Reade, Jr., and dominated the 1890s dime novel market. Its stories highlighted robots, aircraft, submarines, armored vehicles, and steam-powered vehicles. Like Verne, Senarens heralded the use of aluminum and steel. Senarens built a huge market and set the foundation for Victor Appleton's Tom Swift series and the Roy Rockwood Great Marvel series. Today's collectors of Senarens's dime novels find only frustration. The cheap newsprint used has led to disintegration of most books.

There was also an adult science fiction effort based on Edison by Auguste deVilliers de L'Isle-Adam. Villiers, a French writer, was a contemporary of Jules Verne. Villiers drew heavily on the legend like newspaper reports on Edison. Villiers modeled his scientific romances after H. G. Wells, who focused more on the potential for evil through science. This theme in science fiction was uniquely European, whereas the American themes in this genre hailed science as good. Villiers's *L'Ève Future* (1878) played on the ideas of "The Sorcerer of Menlo Park" and "The Magician of the Century." Villiers's fictional Edison combined his many electrical inventions to create an android. Villiers's Edison was "a descendent of both Faust and

Dr. Frankenstein, and like the subject of Mary Shelley's story, his creation of an artificial life form becomes the vehicle for Villiers to meditate on the nature of humanity and the application of technological solutions to human problems."[5] The Villiers novel never really equaled any of Verne's or Wells's, but it illustrated that the Victorian love of legendary scientists was a two-way interaction between reality and fiction. More proof of this comes from Edison notes in which he predicted future events by the use and interference of science.[6] Edison predicted, like Verne, the possibility of a meteor hitting the earth, aerial navigation, huge canal building, and many others. Edison's creativity naturally led him to speculate on the future.

The genre of scientific romance was not limited to Verne, Wells, and children's writers, but attracted such later Victorian writers as Jack London and Rudyard Kipling. Kipling in particular centered a number of short stories around the love of and fascination with machines. Kipling biographer John Brunner noted, "It is one of RK's gifts that he sensed the counterpart of magic in machinery. He became, in a sense the poet of the machine age."[7] While H. G. Wells worried about the effect of machines on man's soul in *The Lord of the Dynamos,* Kipling looked at the soul of machines in *The Ship That Found Itself* and *.007.* Kipling even predicted radio in his story-"Wireless." The work of Kipling illustrated the depth of interest in machines and science throughout Victorian society. Another later Victorian was Edward Bellamy (1850–1898), who predicted radio, motion pictures, television, pedestrian malls and credit cards in his 1888 novel, *Looking Backward.*

The Victorian era birthed a class of dreamer/inventor who combined the fantasy of the romance writers with the skills of inventors. This class of dreamer/inventor can be compared to Leonardo da Vinci. One of these was Charles Babbage, the father of modern computers; Babbage was a skilled machinist and inventor. Yet the "invention" he is best known for, the computer, was never operational in his lifetime. Another dreamer, who even inspired the Victorian science fiction writers, was Etienne Gaspard Robertson. Robertson designed a unique balloon in 1803 that was a basis for the great balloon adventures of Edgar Allan Poe, Jules Verne, and Mark Twain. Robertson designed and hoped to build a balloon capable of carrying 150,000 pounds. He made his proposal to 60 scientists and engineers, hoping to launch a scientific expedition of the world. His balloon was to be called the Minerva. The Minerva's design incorporated conference rooms, an exercise room, church, dining room, cabins for passengers, and cannons. He even added a huge keg of beer underneath to supply the passengers for a two-month journey. Like Babbage's computer, the Minerva would motivate a generation of Victorian inventors.

4

Victorian Iron — The Foundation of the Age

No metal captured the imagination of the Victorian mind more than iron. Iron dominated Victorian art and engineering and ultimately defined its society. Iron was well-known for centuries prior to the 1800s, but it was the Victorians that birthed the real Iron Age. The Victorians built lofty blast furnaces to release iron from its ore. This element, whose origin is in the fiery supernovas of the universe, was the heart of Victorian metallurgy. Iron first revolutionized the military and then supplied the muscle to support the growth of the railroads, ships, bridges and buildings. The many iron monuments of the Victorians still stand today. The Eiffel Tower, built for the International Exhibition of Paris in 1889, was one of those monuments. The furnaces themselves became monuments of industry with a quest for larger and bigger. The press, lacking the phenomena of sports competition, focused on the battle of machines and industrial dreams. There were prodigious tonnage battles like those between the Lucy and Isabella furnaces in Pittsburgh. By the end of the Victorian era, there was anxiety about running out of this elemental industrial blood of iron ore. "Iron hunger" was one of the major topics at scientific conventions, such as the 1910 International Geological Congress in Stockholm.

The term iron requires some explanation. The Victorian evolution of "iron" started with cast iron, then wrought iron, and finally steel. Iron is a chemical element that has a basic atomic purity. Pure iron is soft, malleable, and difficult to manufacture. Elemental iron has almost no application other than medical. The Victorians used mixtures of wine and iron as a tonic for many ailments. The Victorians were familiar with three ferrous (Latin for iron) materials of which iron is the base material. The basic product of early iron furnaces, such as Catalan forges, charcoal furnaces and today's blast furnaces is cast iron. Cast iron is the product of reducing iron ore (oxide) with carbon. Cast iron varies from elemental iron in that it is high in residual carbon (around 3 percent). Products from this process were cast into shapes such as fire plates, stoves, pans, cannons, and art forms. Cast iron, because of its carbon, does not bend readily, but breaks. It is not malleable, as carbon-free pure iron is. Still, it *is* a strong material for building and artillery. Cast iron has what engineers call compressive strength, which makes it the ideal base for heavy machinery. Another unique property is that it absorbs vibration, another plus for machinery bases. Compressive strength and vibration absorption made it the ideal material to replace wooden rails for steam locomotives. Its heat conductivity made it popular for frying pans, stoves, and other cooking applications. Actually these applications are related to a peculiar property that allows cast iron to evenly distribute heat. This property prevents hot spots and allows for even cooking, which accounts for the continued popularity of cast-iron frying pans. Cast iron is naturally high in silicon as well, which makes the molten iron very fluid. The fluidity allows for intricate shapes to be cast. Another attribute of the carbon in cast iron is that it facilitates machining. The excess carbon in the cast iron is in the form of graphite, which acts as a machining lubricant and chip maker.

As you remove carbon from iron, the resulting material is more malleable and workable. The remelting of cast-iron pigs at a foundry causes a decrease in carbon level, which results in early versions of malleable iron, malleable iron being less brittle than cast iron direct from the blast furnace. Malleable iron today is a special type of cast iron, but malleable iron as noted in the Victorian literature was really a poor grade of steel. As you remove carbon, strength decreases. As you approach a carbon level under 1 percent, iron becomes steel because of the change in properties. Steel optimizes the increase in malleability and the decrease in strength. Steel is the ideal of ferrous metallurgy. As you approach elementally pure iron, you have an engineering material known as wrought iron. Wrought iron is soft and easily shaped. The Victorians learned first through the hot working of cast iron to produce wrought iron before they discovered the intermediate carbon product of steel.

The basic production of iron from ore is described in Jules Verne's *The Mysterious Island* (1875):

> For, strictly speaking, the Catalan method requires kilns and crucibles, in which the ore and coal are arranged in alternating layers before they are exposed to the heat of the fire. But Cyrus Smith wanted to avoid any unnecessary steps, and so planned simply to make a cube-shaped pile of ore and coal, into the center of which he would direct the flow of air from his bellows. This must have been the technique used by Tubalcain and the first metallurgists of the inhabited world. And if it had worked for Adam's grandchildren, if it was still successfully practiced in lands rich in ore and fuel, it could not fail to work for the colonists of Lincoln Island.
>
> After the ore, the coal was gathered from exposed deposits they had found a

small distance away. They broke bits of ore into small pieces and scraped away any visible impurities from the surface. Next the coal and ore were arranged in successive layers to form a large pile just as a charcoal maker does with the wood he wishes to carbonize. In that way under the influence of the air injected by the bellows, the coal would be transformed in carbonic acid, and then into carbon monoxide, whose role is to reduce the iron oxide — to rob it of its oxygen, in other words.

Verne described an advanced technology of the 1850s using coal. The Victorian age started with charcoal made from wood as the carbon source in the furnace. Burning hardwoods such as oak, chestnut and hickory produced charcoal. While coal was the primary source of fuel for iron making in England by 1840, it wasn't until 1850 that its use started in America. Charcoal actually is a superior fuel for iron making because its lacks impurities, such as sulfur. Its major shortcoming was that it required an acre of hardwood to run a furnace for one day. By 1750, England was already running short on hardwood forests. England also had laws reducing the use of hardwood except for critical applications in shipbuilding. Even the great forests of America could not have supplied the iron industry for more than a few decades. In fact, northern Michigan, which was rich in iron ore but lacked coal, did start charcoal iron making in 1860 but lumbered out the area in less than two decades. It is said of charcoal furnaces that "each week the small ironworks consumed a football field of virgin forest to produce a few tons of pig iron." A typical charcoal furnace in 1800 used about 5,000 to 6,000 cords of wood a year, which translates to 240 acres of forest a year.[1] The charcoal iron furnace has long since disappeared in the United States, but today the Brazilian iron-making industry is using charcoal as fuel, and this is the major reason for the loss of the rain forests.

The product of a charcoal iron furnace was "pig" iron, often just called cast iron. These pigs were made by running molten (liquid) iron into dirt trenches on the furnace floor. There was a main runner coming from the tap hole of the furnace. Then side runners were made in a treelike structure. Because the side runners looked like baby suckling pigs feeding off a mother pig, the term pig iron came into common usage. The pig was a rectangular block running 12 to 24 inches and 6 inches by 2 inches on the end. The liquid pig iron was a product of charcoal taking away the oxygen in iron ore (known as smelting or reducing) to produce iron. This iron, however, was not pure but a mixture of carbon and iron. Pig iron contained about 3 percent carbon and 1 percent silicon. The "pigs" were about 10 to 100 pounds to allow easy handling.

Pig iron was an intermediate product. It was sold to foundries to remelt and cast into shapes, to puddlers to make wrought iron, and eventually to steel makers. Cast iron generally referred to shaped products being produced in foundries by re-melting pig iron. Cast iron was a much-beloved engineering material of the Victorians. Applications included stoves, machinery, broilers, locomotives, lawn furniture, bridge beams, frying pans, and even cast buildings. The heavy richness of cast iron seemed to capture the spirit of the age. The technology of commercial cast iron goes back to James Darby's innovation of the use of coke in blast furnaces in the 1700s. Darby's Coalbrookdale Ironworks built the first cast-iron bridge in 1777. It should be noted that while Darby's famous bridge dominates history books, the first attempted iron bridge was at Lyons, France, in 1755.[2] Cast iron had limitations because of the brittleness and a low tensile strength inherent in the high carbon. For centuries, sword makers and smiths had used cast iron as a starting point to produce wrought iron and steel. This process was known as "heat and beat." The smith would heat cast iron, burning out carbon, and then hammer the softer product to burn out even more carbon. Ultimately, this slow routine produced low-carbon wrought iron. Wrought iron is a ductile product that can be hammered or mechanically worked into shapes. It had good tensile as well as compressive strength.

In 1830, about 85 percent of all pig iron produced was destined to be used in the making of wrought iron.[3] Wrought iron was made in a process known as puddling, which involves reheating pig iron in a puddling furnace and "working" it to burn out carbon. Puddling offered a quicker and easier route to remove carbon for cast iron than heating and beating. Historian James Howard Bridge writes that puddling required the iron be: "reduced to liquid form and boiled and stirred about until most of the impurities were driven off [primarily carbon — the chemistry of the process was not well understood at the time]. When the bubbling mass thickened and assumed a pasty consistency, the puddler passed a long bar through a small opening in the furnace door, and rolled the paste into a ball. This ball was then withdrawn and carried, dripping with liquid fire, to a queer arrangement of big wheels which crushed and rolled the ball over and over, squeezing out all sorts of useless stuff and further solidifying the mass.... The ball was then re-heated, and passed under hammers and through rollers; and the kneading it thus repeatedly underwent gave it the fibrous quality of wrought iron. When it had been finished into bars it was ready for the market."[4] Of the remaining 15 percent of charcoal-furnace iron production not made into wrought iron, most of it was in casting heating stove production, which also required a remelt process at a foundry. Very little product was cast directly from the charcoal or blast furnace operation. In 1850, only 50 percent of cast pig iron was going to wrought iron as more was being applied to cast iron, steel and malleable iron production. By 1880 more than 85 percent of blast furnace production was going to steel making, with less than 5 percent to wrought iron.

While the English had discovered the furnace technology to switch from charcoal to coal by the late 1780s, the new technology offered little quality and cost advantages in America. It did greatly increase the productivity of an individual furnace and did not require plentiful wood, but American just built more furnaces instead of switching to coke. Coal did address the declining resource of hardwood forests. The British for almost two centuries had dealt with a struggle for the hardwood resources between iron making and shipbuilding, which required good oak. In 19th-century America, as in Brazil today, there was no economic motivation to move to coal or coke. Plentiful hardwood forests seemed inexhaustible in America. However, by 1835 at local iron producing areas in the East, hardwood forests were disappearing. In 1835, the Franklin Institute of Philadelphia offered a gold medal "to the person who shall manufacture in the United States the greatest quantity of iron from the ore during the year, using no other fuel than bituminous coal or coke, the quantity not to be less than twenty tons." The medal was never awarded, but it was typical of the competitive nature of science in the Victorian era. By 1850, the British industry was all coke furnaces while the United States remained mostly charcoal iron. The end for charcoal was coming. A small iron industry in northern Michigan lumbered out the northern thumb and the forests of New England in a few decades. The real bottleneck to the use of iron products was the limited tonnages being produced in blast furnaces. Charcoal furnaces in particular were limited to a stack height of 40 feet and a weekly production of 150 tons a week. British coke furnaces could easily double that production with their higher stacks.

This question of blast furnace fuel is a complex issue of economics, geology, mineralogy and metallurgy. There are two board classes of coal: anthracite, a very hard coal known as "natural coke," and bituminous coal, a soft, gaseous coal that required coking to be used for iron-making furnaces. Coking is an intermediate burning of coal that required special coking ovens to produce. Bituminous coal was found in a major concentration in western Pennsylvania, West Virginia and Maryland. Anthracite coal deposits resided mainly in the eastern mountains of the United States. Since the East was feeling the effects of the lumbering out of hardwoods, the switch to anthracite became a natural progression. This eastern movement to anthracite coal began in 1840. By 1854, when the American Iron and Steel Institute began keeping statistics, 45 percent of pig iron production came from anthracite coal, reflecting the eastern domination of the iron industry. About 7.5 percent was from bituminous coke with the balance, some 46 percent from charcoal. In addition, a protective tariff on nondomestic iron passed in 1842 allowed the Americans to continue to use less profitable charcoal. Blacksmiths, wrought-iron makers and steel makers also favored charcoal iron because of its purity. More than 150 years after the disappearance of American charcoal, there remains a small market for charcoal pig because of its purity. The purity and quality advantage arose from the lack of sulfur and other impurities in charcoal that were often high in coke made from bituminous coal. Slowly, the increased demand for pig iron, the decline of hardwood forests, the shift of the industry to the western Pennsylvania coal regions, and the need for larger integrated furnaces forced the movement to bituminous coke. Certain hardwood-rich areas, such as northern Michigan and southern Ohio, bucked the trend. The Hocking Hills area of southern Ohio had a prosperous charcoal iron industry from the 1860s to the 1880s. Some of these "high tech" charcoal furnaces still stand today for visitors to view. In any event, by 1890 the charcoal iron industry was all but gone, with the exception of a brief recall of these old furnaces to meet the demand for pig iron during World War I. Coke ultimately allowed for larger and hotter furnaces, which in turn boosted productivity and reduced prices.

In 1828, a very important but historically forgotten invention appeared in England. It was the use of hot-air blasts into the iron-making furnace. The inventor of this process was James B. Neilson. Neilson found that preheating air prior to air-blasting the furnace doubled the furnace output with no increase in fuel! This resulted in amazing increases in productivity, unequaled by future technological advances of the industry. Furthermore, the use of hot-air blasts allowed for higher heat and ultimately the use of lower-grade ores. This alone opened up the lower-grade Scottish ore and propelled Scotland into a major iron exporter. Between 1829 and 1845 the annual production of pig iron rose from 29,000 tons to 475,000 tons. It should be noted that coal as a fuel was needed for Neilson to achieve these results; thus they could not be duplicated in American charcoal furnaces. Charcoal furnaces could not hold the tall stacks of ore and fuel. Charcoal had always produced a lower-carbon product than coke or coal blast furnaces. The immediate impact was a large decrease in the cost of iron. High-quality British iron flooded America because of its cost advantage.

The British maintained the advantage for almost thirty years. Another advantage came as a product of serendipity. The hotter blast removed more carbon from the cast iron, producing a more malleable form of iron, as well as wrought iron via rolling. Sending Neilson's hot-blast pig to foundries helped produce an even lower-carbon product. There is even evidence that some furnace managers focused the hot blast on the liquid pig iron at the bottom of the furnace, producing a type of steel or "malleable" iron. This type of stirring the liquid pig iron in the charcoal furnace is equivalent to the Bessemer converter stirring. The malleable iron was far less ductile and malleable than high-grade steel, but such steel in the first half of the 19th century was extremely expensive. It should be noted that Neilson did not invent what we know today as "malleable iron." Neilson's malleable iron was really a high-carbon steel product. Several inventors had produced malleable iron decades earlier, but it required remelting and long heating of cast iron. Neilson's furnaces and experiments laid the groundwork for future steel makers William Kelly and Henry Bessemer. In fact, we know Kelly was familiar with hot-blast malleable iron from his work with charcoal furnaces, and Bessemer learned and studied such hot-blast experiments via scientific conferences. Neilson was able to produce a lower-carbon product out of the blast furnace. Cast-iron water pipes emerged as one new application in England. The water pipes of the big English cities prior to 1840 were wooden logs with holes drilled in them. These wooden pipes were jointed together by wrought iron bands. Malleable cast iron remains today the preferred material for water pipes (although its much improved). Neilson's cast iron was cheaper than wood and outlasted wood by a factor of four. Imported "malleable" cast iron was being used in Philadelphia by 1830. Many cast-iron pipes around today have service times of more than 125 years.

Cast iron captured the imagination of a little-known American building engineer, James Bogardus. Bogardus in 1840 suggested the use of cast iron in buildings. Cast iron was fire-resistant and could be cast into detailed columns and panels. In 1849, Bogardus erected the first self-supporting building with exterior iron walls—the Edgar Laing stores. Bogardus went on to develop prefabricated cast-iron façades. The use of cast iron became very popular throughout America because of its fire resistance. Major cities throughout the United States adopted cast iron. The Cast Iron Historic District in New York City has 139 iron-façade buildings, but most major American cities had examples of cast iron in architecture. The largest example of cast iron columns was Chicago's sixteen-story Manhattan Building. In 1890, the Manhattan Building was the world's tallest skyscraper. Cast iron also became popular in decorative steps, railings, fences, pedestrian bridges, window frames, and shutters. Central Park in New York still has an excellent collection of pedestrian cast-iron bridges.

While America developed cast-iron buildings, England perfected the cast-iron bridge. The design was based on a simple arch used by the Romans to build stone bridges. The arch design puts the material in compression, which favors cast iron. The Coalbrookdale Bridge was famous throughout the world, becoming a sort of mecca for visits by Victorian engineers. Only a few of these arch-type bridges were built in the United States. The oldest in the United States (1839) is over Dunlaps Creek in Brownsville, Pennsylvania. It was designed by Captain Richard Delafield of the Army Corps of Engineers as part of the National Road. It survives today and still carries traffic.

Another unusual application of Neilson's "malleable" iron was the use of malleable iron chain for ship anchors. For centuries heavy metal anchors were carried by thick ropes. These ropes were subject to decay from seawater, animals and bacteria. To preserve them required almost constant cleaning by sailors. Neilson's cast iron changed all that and made the sailor's life a lot easier. Iron products also started to replace wooden railroad rails with Neilson's cheaper product. Wrought iron, however, would soon dominate the rail market. Neilson set up a combination of hot-blast coke iron and rolling that produced wrought iron in addition to intermediate products such as malleable iron. Neilson's "malleable" iron could be used for railroad rails, and the process of replacing wood started. Neilson's productivity and efficiency improvements further reduced the cost and increased the application of wrought iron and steel. His coke-fueled, hot-blast furnace was a necessary step for the development of the future Bessemer steel industry of 1870s. With these technogical advancements, the groundwork was set for steel, but prior to that, reduced pig iron prices would make wrought iron the material of choice in the middle Victorian era.

Another typical Victorian element of the development of iron furnaces was the quest for bigger and better. In particular, furnaces took headlines in iron-making towns such as Pittsburgh and Essen. One of the most famous of these competitive battles took place in Pittsburgh during the 1870s. It was between the Lucy Furnace of Andrew Carnegie and the Isabella Furnace of some Pittsburgh competitors. The furnaces were of equal size and produced an average of 50 tons of iron per day. The competition started with the local papers reporting new daily records. When the Lucy Furnace produced 100 tons in one day, the international and trade journals started to give weekly reports. In England the creditability of such records were even questioned. Both these furnaces were, however, coke furnaces, like those in England. Their successes helped speed the movement to coke in the United States. This monumental tonnage battle went on for seven years. One historian commented, "When the eyes of the American people are lifted from the kindergarten romances of myth and fiction to grandly moving epics of industry that give distinctive value to the history of our republic, their story will become a national heritage."[5] This battle birthed a number of furnace improvements that would become the foundation of the Carnegie steel empire. Competition actually moved Lucy Furnace to become the first to hire a chemist to help find new ways to apply science to the competition. Competition was a fundamental element in the Victorian advances in iron making. By 1880, furnace records were recorded in the press like baseball scores today.

Initially the demand for low-carbon wrought iron was in nail making and sheet metal for product such as tin plate. Tin plate in particular was a commercial success of the Victorians. Tin plate was produced by hot-dipping sheet iron in molten tin. Tin had been recognized since the Romans as an inert metal coating, inert meaning that the metal is not oxidized or affected by natural acids. (Gold, tin, platinum, and rhodium are examples of inert metals.) The Romans coated copper cookware with tin to eliminate the metal taste from copper oxides. Tin-plating probably goes back to 17th-century Germany. The application of tin plate was limited, however, because of the short supply of wrought iron. An obscure Frenchman, Nicholas Appert, gets credit for the "invention" of tin plate even through Peter Durand appeared to be following the stream of development leading to independent patents in 1810. The French claim of invention seems more credible. Appert was known to have supplied test tin cans of food for Napoleon's army. He preserved partridges, various vegetables and gravies. After five months of storage, the food retained its freshness. Appert was awarded a French prize proposed in 1795 by the French Directory to find an effective means of preserving food. The award was given by Napoleon, who saw its importance in long strategic military operations. Napoleon proved its success in his long march to Moscow. Peter Durand rallied British resources to meet the challenge and became a major European producer.

In 1815, Russian explorer Otto von Kotzebue used the English "tin boxes" of Durand in his search for the Northwest Passage. Sir William Edward Parry also used "tin boxes" in his 1819 search for the Northwest Passage. These new tin boxes did have some inherent dangers because of the lead/tin solder used. Recent studies of the famous lost Northwest exploration of John Franklin in 1845 suggest the solder was a source of lead poisoning. Researchers at the University of Alberta were allowed in 1988 to exhume bodies from Franklin's failed effort. The findings showed lead poisoning exacerbated by scurvy. Testing of the remaining cans showed poor design allowing lead to be leached into the food. Victorian doctors were, however, completely ignorant of the medical side effects of lead.

In the United States, a third inventor was actually producing tin cans in 1812. Thomas Kensett set up a factory in New York to produce hermetically sealed glass jars as well as tins. He was granted an American patent in 1825. Still, the real boom in tinplate didn't happen until the 1850s and 1860s. The demand for preserved food increased with the 1849 gold rush and westward movement. The Civil War further stimulated demand in the 1860s. This demand was coupled with the increase in the puddling process to produce wrought iron. The Civil War soldiers became accustomed to the use of tins to preserve food. After the Civil War the demand for tinplate accelerated until 1890. Almost all of the tinplate for American canneries came from Welsh mills. It wasn't until American steel rose to prominence that America became a major producer of tinplate in the 1890s.

Before the market for tinplate fully blossomed, wrought iron became the material of choice, replacing cast iron as a structural material. Wrought iron could handle mechanical stress better than cast iron. Thomas Telford, a British engineer, compared the engineering properties of cast iron and wrought iron for bridge building as early as 1811. Cast iron results in strong, rigid structures, but the structure is susceptible to breakage from bending and twisting. Also, cast iron is more difficult to work with. Telford's Menai Straits Bridge in Wales started out as a cast iron arch bridge, but the occurrence of gales in the area concerned him enough to go to a suspension bridge. Telford did endless experiments before and during the seven years of construction. The suspension in this case was wrought-iron eye bars tied into chains. The ability to handle twisting and pulling made it the ideal material for bridges. As the railroads grew, the need for stronger bridges grew also. The earliest cast iron bridges were, as noted, arch-type bridges. As early as 1820, Robert Stephenson in England was experimenting with new designs. Stephenson's first innovation was the cast-iron beam, which allowed him to combine arch and suspension designs. Stephenson moved to wrought iron as puddling made it available in large quantities. His first use of wrought iron for a bridge was the Gaunless Viaduct in 1825. It was a combination of cast-iron posts and wrought-iron rods. In 1829, Stephenson used wrought-iron beams to form an arch bridge known as the River Tyne Bridge. In 1850, Stephenson and William Fairbairn opened all terrain types to the railroads by the use of bridges and tunnels. The wrought-iron design used the idea of reinforced rectangular "tubes" in the form of a suspension bridge. This Britannia Bridge crossed the Menai Straits in Wales with four spans of 459 feet each. This bridge was probably heavily overdesigned, which contributed to its longevity. Stephenson was sensitive because in 1848, one of his composite iron bridges failed (Dee River) due to resonance. Resonance causes the metal to fail by harmonic vibration developed by winds or rhythmic marching. The Britannia Bridge pioneered the new technology of pneumatic riveting invented by William Fairbairn.

In America, bridge-building experiments took a different direction. As early as 1816, Erskine Hazard had built a successful wire suspension bridge across the Schuylkill River at Philadelphia, but it was merely a temporary footbridge. In the 1830s, France led the technology of suspension footbridges, but no one had built anything capable of carrying the loads found in railroad construction. John A. Roebling in the 1840s suggested the possibility of a suspension railroad bridge. This belief was based on the cumulative technology of a number of innovative streams that Roebling had studied. In particular the experiments of Telford in the 1820s pioneered iron wire. Robert Stephenson opposed even Roebling's suggestion. The key to Roebling's design was the use of wrought-iron wire made into rope. Roebling started in the 1840s to develop iron rope based on some earlier work in Germany. The Germans had successfully drawn iron wire and then twisted seven wires into rope, but only a very small quantity had been made. In the early 1840s, Roebling built the necessary equipment to produce large quantities of iron rope. The first application of his iron rope was in canal portages. Hemp ropes had proved unreliable and short-lived. Roebling's iron rope had many times the strength at a much smaller diameter. In addition, iron rope had a much longer life. This success led him to believe iron rope could be used to suspend a bridge.

Roebling began such a project in 1851, with a railroad bridge at Niagara Falls. Its main span was 821 feet and stood 240 feet above the water. The piers were anchored on each side of the gap. The bridge was suspended with four 10-inch cables that used 3,640 wrought-iron wires each. Robert Stephenson declared, "If your Niagara Bridge succeeds, mine have been magnificent blunders."[6] Stephenson had good reason to be pessimistic, since in 1850 the collapse of the Bassie-Chaine Suspension Bridge in France had resulted in a European moratorium. In fact, Roebling succeeded, and the bridge was the first suspension railroad bridge. A little later Roebling built the Sixth Street Bridge in Pittsburgh with iron rope, which did show some wear after 40 years and required rework. The suspension design would become a standard with the coming of steel wire and the building of the Brooklyn Bridge by his son, Washington Roebling.

Another breakthrough in wrought-iron bridge building came in the 1860s with the invention of the cantilever method. The inventor was Isambard Kingdom Brunel (the future designer of the *Great Eastern*). Prior to the cantilever approach, arch bridges were built with timber support for the span as it was constructed. The cantilever approach starts with the building of two half arches on each pier. These half arches are loosely tied to the piers with temporary rods. Finally a center can be floated to position and raised to then close the arch. The principle greatly improved the ease and speed of construction. This technique was applied in the building of the first steel bridge at St. Louis in 1874. Steel's natural strength allowed the cantilever bridge design to gain in popularity. Steel has an unusual combination of metallurgical characteristics: flexibility, strength, and ductility. Wrought iron has the flexibility and ductility but lacks the strength that can be developed in steel.

Another application of wrought iron was in buildings. In the 1850s, Peter Cooper rolled the first wrought-iron beams to be used in buildings. One the first two buildings, the Cooper Union Building (1853), is still standing. These structural beams were known as I-beams because of their shape in a cross-section view. Cooper erected a rolling mill in Trenton, New Jersey, to produce I-beams using puddled iron. Rolled wrought iron opened the possibility of the modern skyscrapers, but six stories was considered the limit people would walk prior to elevators.[7] Thus, even with the new strength of iron, building heights remained limited until elevators. The invention of the elevator safety device by Elishia Otis in 1854 allowed for the building of nine to ten stories. Still, such higher buildings used masonry to carry the weight until 1885. In 1885, Andrew Carnegie combined six stories of wrought iron and six stories of steel beams in the Home Insurance Building.

5

William Kelly — The Inventor of the Bessemer Process

In the 1850s a new technology was on the horizon that would transform Braddock's Field (the future site of Andrew Carnegie's first steel mill) outside of Pittsburgh. That sleepy village had known little activity since the defeat of General Braddock and a young colonel, George Washington, during the French and Indian War. William Kelly invented this new technology known to the Irish of Braddock as the Kelly Pneumatic Process but to the world as the Bessemer Process. That process would light the night skies of Braddock and the world for more than 75 years. The Bessemer Process was a steel-making process. Bessemer steel fueled the Victorian railroad boom, which in turn fueled a steel boom. It was so-called Bessemer steel that made Andrew Carnegie one of the richest men ever to live. By 1895, Bessemer steel production accounted for more than 60 percent of the gross national product.[1] The wealth generated by Bessemer steel production turned America into the premier industrial nation in the world. Bessemer steel created hundreds of millionaires throughout the United States alone. A group of Carnegie executives known as the "Boys of Braddock" controlled 6 out of 10 of America's largest industries from 1900 to 1930. It was the profits from Bessemer steel that gave the world more than 2,000 libraries, as well as museums, concert halls and art exhibitions. The inventor of this revolutionary technology was William Kelly, yet the name Bessemer is memorialized in street maps of the steel regions throughout the United States.

The pneumatic steel-making process (commonly now referred to as the Bessemer Process) was invented by the Irish Pittsburgh steel maker William Kelly. Kelly was at least 10 years ahead of Sir Henry Bessemer in the development of the process. But Bessemer, a seasoned inventor, got to the patent office first. The victory went to the first person with patent rights, though the court battles that followed favored a reversal of the original decision.[2] The Victorian love of a hero and industrial winner left room for only one "inventor," and the public was eager to declare a winner. The pneumatic steel-making process in hindsight was like most of the great Victorian inventions, a triumph of streams of scientific advancement rather than an individual event or person. In 1875, Andrew Carnegie built his first Bessemer behemoth plant at Braddock, Pennsylvania. Carnegie had purchased the patent rights to the Bessemer Process, but the majority Irish workforce refused to call this pneumatic steel-making process the Bessemer Process. For more than a century, a core of Irish steel-makers would surrender such an honor only to fellow Irishman Kelly. Even today, steel-makers like me see it as the Kelly Pneumatic Process.

Again, a little metallurgy is needed to proceed. Steel as well as wrought iron can be made from pig iron by reducing the carbon content from 3 percent in pig iron and 1 percent or less for steel to .2 percent for wrought iron. One of the results of reducing iron oxide with carbon is that the resulting pig or cast iron is high in residual carbon. Oxygen, heat, and working of the iron are the means to burn out this residual carbon and produce steel. The "heat-and-beat" method used to remove carbon was the earliest method used for steel production. Its roots go back to the great sword-producing areas of Toledo, Spain, Damascus, and Japan. The heating and beating exposed the carbon in the iron to air, which burned out the carbon. Heating and beating was a slow, labor-intensive process. It was the Catalan forge method, where a pasty pig iron came out of the furnace and was then hammered into wrought iron or steel.

The Victorians had improved on the Catalan method with the introduction of puddling. Puddling was actually patented in 1784 by Henry Cort. As we have seen, puddling evolved from the process used to produce wrought iron by eliminating carbon. The process took cold pig iron bars and reheated them in a reverberatory furnace (invented in the 1600s). A reverberatory furnace separates the fuel from the pig iron in the furnace and uses the flames and hot air to heat the iron. This separation stops the pick-up of more carbon from the fuel. The mushy paste created in the furnace was "stirred" or puddled by a puddler. This heating and puddling exposed the pig iron to oxygen in the air, which burned out the carbon inherent in the pig iron. Initially, puddling created another immediate product for foundries—"malleable" cast iron, which was a crude type of steel. The goal of puddling, however, was a very low-carbon iron. The puddling process was designed to totally eliminate carbon, producing a wrought iron, although techniques to make the higher-carbon product of steel were in place. The majority of puddled product was wrought iron by process design. Wrought iron represented 100 percent of the process until about 1850, when steel started to be made using puddling. To produce steel from puddled iron actually required the addition of carbon into reheated wrought iron. The

Victorians' improved on the heating and furnace design so that even lower-carbon steel could be produced in volume.

Another 18th-century steel-making process the Victorians improved on was the crucible process of Benjamin Huntsman. Huntsman rediscovered an ancient Chinese method of making steel. Pig iron could be put in a ceramic crucible and heated for days. The long heating caused the carbon to eventually burn off. There was, however, a limit to carbon removal because the very crucibles were made of graphitic carbon. Huntsman had developed the process around 1740 in his Sheffield Works, which produced quality steel. Huntsman and his steel-making process would of course make Sheffield steel cutlery world famous. He kept the process secret for most of the 1700s. Huntsman swore his workers to secrecy and did all steel making at night to further prevent spying. Spies eventually did get in, and rival companies appeared in Sheffield, but England retained the full secret until the 19th century. Crucible steel found new markets, one of which was cast-steel dies and rolls for the manufacture of gold coins. Crucible-steel bayonets were also gaining popularity in Prussia and France. The problem was that in 1800, England had a steel monopoly and controlled Europe's supply.

To break the steel-making monopoly of England became a passion of Napoleon, and in true Victorian methodology, he offered a reward — 4,000 thalers to the first manufacturer on the continent who could turn out cast steel of British quality. This competition led to the formation of the Krupp Cast Steel Factory in Germany. By the late 1830s, Alfred Krupp was producing cast crucible steel, but it still lacked the quality of British cast steel. Krupp took a grand tour of England and Europe to learn steel making. Krupp biographer Norbert Muhlen noted his quest: "Rather than make steel, it now became his ambition to make 'the best steel,' and he pursued his goal with fanaticism. Krupp used spies to help fully understand the Sheffield steel making process. A crucible steel cylinder to roll gold which he delivered to a manufacturer at Valbom, Portugal, in 1840, was still in use and in perfect shape as late as 1958."[3] It should be noted that the average life gold coining rolls prior to this was 4 to 5 years!

The real prize of Alfred Krupp was the winning of the gold medal at the 1851 Great Exhibition for his cast-steel cannon. A cast-steel cannon was a first for the world and hastened the replacement of Napoleonic bronze and cast iron. In addition, Krupp showed the largest cast-steel block, defeating the host British effort at the largest steel-casting competition. The arrival of the winning casting and rumors of it were followed closely by the British press. Krupp surprised the British and the Americans, helping to launch even more international competition to produce quality. International competition was a key underlying factor in the overall creativity boom of the

Victorian Metallurgic Age. Removing carbon from pig iron efficiently to produce steel became the focus of many.

The production of steel, and particularly cast steel, stirred the national pride not only of England but also the United States. America was far behind, with fewer than 50 tons of cast steel being produced in 1850. Even wrought-iron production lagged behind, but 22,000 tons were produced for rails that year. In 1852, Krupp started producing steel rails as well as cast-steel wheels for locomotives. It wasn't until 1855 that the first steel rails were rolled in America at Cambria Iron Works in Johnstown, Pennsylvania. Rails were becoming a major market, and many foresaw steel with its superior strength and wear properties overtaking wrought iron. The volume of steel being produced as a replacement for cast iron and wrought iron would strain the crucible process. A new process would be needed to launch the Age of Steel.

William Kelly, as a chemist, understood that air (oxygen) was actually eliminating the carbon in the pig iron. The heat-and-beat approach took cold pig iron, heated it up and then worked the iron in some method, such

as rolling or hammering. The "working" exposed the surface to the air and burned the carbon out. Kelly understood the chemistry, and he produced lower-carbon cast iron, known as "malleable" iron, in the 1840s. He probably got this idea from some of his early work with creative Pennsylvania furnaces, which were blowing air into the bottom of the furnace to make a "malleable iron"–type product. Kelly had extensive furnace-operation and building experience, which would have given him some practical application of the use of air to burn carbon out of iron. He had worked with the application of Neilson's hot blast in charcoal furnaces in the hills of Pennsylvania. Kelly's new process used molten (liquid) pig iron from the blast furnace. This liquid iron was then poured into a "converter," which was like a brick-lined kettle to hold the liquid. Air was then blown through the bottom

Sir Henry Bessemer — meta-inventor and inventor of the pneumatic steel-making process

of the converter by means of a tuyère, a pipe designed for that purpose. The oxygen in the air burnt out the carbon directly. The process produced a fountain of sparks and light that could be seen for miles.

Bessemer himself described it eloquently: "The powerful jets of air spring upward through the fluid mass of metal. The air expanding in volume divides itself into globules, or bursts violently upward, carrying with it some hundredweight of fluid metal, which again falls into the boiling mass below. Every part of the apparatus trembles under the violent agitation thus produced; a roaring flame rushes from the mouth of the vessel, and as the process advances it changes its violet color to orange, and finally to a voluminous pure white flame."[4] Another good description is that of steel historian Stewart Holbrook: "To the layman it looks like the egg of a roc [mythical Arabian bird of enormous size], that fabulous bird which was said to have borne off the biggest elephant in its flight. A roc's egg with one end cut off and gaping. It is a container of brick and riveted steel, twice as tall as the tallest man and supported near its middle on axles. It is set up high on groundwork of brick. Into this caldron goes molten ore, fifty thousand pounds at a time. Through the iron is blown cold air—oxygen forced through the hot metal with the power of a giant's breath. Out of the egg, in good time, comes steel. It is little short of pure magic."[5]

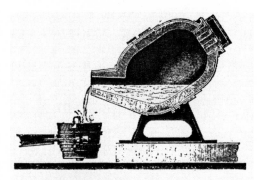

Bessemer converter design from *Scientific American,* 1900

While he probably had the process developed by late 1849, Kelly did not apply for a patent until 1852. Kelly clearly defined the chemistry of the process, which Bessemer used to develop the supporting equipment. Bessemer's engineering expertise allowed English mills to adopt the process far ahead of the Americans. In America, Bessemer beat Kelly to the patent office by a few days. Initially, Bessemer won the legal battle, but as an Englishman, Bessemer never earned the respect of the American Irish.

In 1860, Captain E. B. Ward of Detroit licensed the Kelly Pneumatic

Process and the Bessemer equipment. Daniel Morrell of Johnstown was part of the company that built this first pneumatic steel plant in Wyandotte, Michigan. Morrell had come from the Cambria Iron Company of Johnstown, which was famous for the production of wrought-iron railroad rails. While the Detroit plant was to produce structural steel, the hope was to break into the steel-rail business. The first Kelly (Bessemer) steel was rolled in 1864. The plant used the machinery invented by Henry Bessemer, who had independently "invented" the pneumatic steel process equipment in England. The chemical control of the steel, however, was based on Kelly's work. Kelly held a degree from the University of Pittsburgh, while Bessemer was a generalist and mega-inventor with inventions such as gold paint, sugar presses, postage-cancellation machines, and steel-making process. Both men, however, were driven by the economic rewards of a booming railroad industry. Railroads were just starting to experiment with steel rails in 1867, but the future market was huge. The miles of railroads increased in 1864 from 33,908 miles to 77,740 miles in 1874.[6] That growth would again double by 1884 with the introduction of Bessemer rails. Bessemer steel could carry 20 times the load and last triple the life of iron rails. Such an amazing improvement resulted in huge transportation productivity increases, which further simulated industrial growth.

In 1864, a rival Bessemer rail plant was built in Troy, New York. This plant used the full Bessemer patent and was designed by engineer Alexander L. Hollry. This Troy plant was actually partially owned by the Pennsylvania railroad. The Troy plant would become key to George Westinghouse's use of Bessemer steel in his famous air brake. The selection of Bessemer's patents for use at American steel plants was due to the financial interests of Holley, the chief engineer. Holley had visited Bessemer in England in 1862. He procured the license from Bessemer for America. Holley formed Albany and Rensselaer Iron and Steel as a Bessemer steel construction company. Holley went on to build Cambria Iron's Bessemer plant and then in 1875, a Carnegie behemoth, Edgar Thomson Works. Holley, as the leading Bessemer engineer in the United States, probably won the commercial battle for Bessemer. While Kelly ultimately won the patent battle, he lost the commercial battle. Kelly should be called the inventor of the process, while Bessemer should rightly be called the inventor of the machinery. It was the chemical genius of Kelly that hypothesized the reactions, while Bessemer's equipment energized those reactions. For simplicity, the accepted historical name "Bessemer process" will be used throughout the book. Whatever the name of the process, pneumatic steel making would dominate the railroad industry.

Actually the Cambria Iron Company in Johnstown claims to have rolled the first steel rails for railroad rails in 1855.[7] This was probably an experimental batch of crucible steel for the Pennsylvania Railroad and its manager Edgar Thomson. This rail steel was either made by a special "heat-and-beat" method or was a very limited batch of crucible steel. In any case it was not Bessemer steel. The first trials in 1862 of steel rails by the Pennsylvania Railroad under then-president Edgar Thomson were crucible steel. Cambria would roll the first Bessemer rails in the United States in 1867. The Bessemer ingots were made at a Bessemer plant in Steelton, Pennsylvania, and rolled at Cambria.

Crucible steel's higher carbon content meant more brittle steel than Bessemer steel. The Pennsylvania Railroad learned this the hard way with the breaking of some of the rails in service. Winter temperatures make steels even more brittle, so the winter was particularly hard on these crucible-steel rails. The Pennsylvania Railroad had been unaware of this cold-weather embrittlement. The Pennsylvania Railroad studies, however, did note the high wear on these first experimental steel rails, but believed the future would be steel if its toughness could be improved. Another problem of crucible steel was that it cost almost ten times as much as Bessemer steel. In addition, large castings, as we saw with Krupp's great 1851 ingot, required either simultaneous pouring of many crucibles with German precision or the fictional pen of Jules Verne. Such sequenced pouring also resulted in a mixed chemistry since each crucible batch had a specific composition. Such a mixed or layered chemistry is extremely detrimental to the performance of steel. Crucible steel also, because of its graphite crucibles, was naturally higher in carbon and lacked the ductility needed for rail production.

Blowing air through the molten pig iron in a vessel made lower-carbon steel. The Bessemer or Kelly converter vomited a steady stream of sparks in the process of eliminating carbon. The night skies of Pittsburgh were legendary for their brightness and orange color from Bessemer converters. Carnegie historian James Howard Bridge wrote, "The chemical changes accompanying this gorgeous display are equally beautiful." The fountain of sparks became a snow of iron dust throughout the Monongahela Valley. Many Pittsburghers remember well the need to sweep this dust off porches every morning in the days of Bessemer converters. The Bessemer would also put a heavy smell of rotten eggs throughout the valley due to sulfur. In locations like Bessemer Terrace, on the hill above Edgar Thomson, the dust could be a quarter-inch thick overnight!

Another great description of what the environment looked like comes from a young Charles Schwab, future Carnegie partner and first president of U.S. Steel. Schwab lived as a boy 20 miles northeast of Johnstown's Cambria plant. "Along toward dusk tongues of flame would shoot up in the pall around Johnstown," Schwab recalled. "When some furnace door was opened the evening turned red. A boy watching from the rim of the hills had a vast arena before him, a place of vague forms, great labors, and dancing fires. And the murk always present, the smell of the foundry. It gets into your hair, your clothes, even your blood."[8] This is a description that Pittsburghers of the 1940s can well remember. The orange nights represented an economic indicator for the nation. The smell of sulfur was to any steel town the smell of prosperity.

Eddyville, Kentucky, is the birthplace of Kelly's so-called Bessemer Process. Kelly, a Pittsburgh native, had been trained at what is now the University of Pittsburgh in metallurgy. Kelly had worked for some of the most inventive charcoal iron furnaces of Western Pennsylvania. In particular, Kelly had seen the use of air blowing to produce "malleable" iron in charcoal furnaces. Kelly moved to Kentucky to build and work out his pneumatic process in the 1840s, seven years ahead of Bessemer. Bessemer, however, was the first to patent the process. Bessemer was granted a British patent on February 12, 1856 (No. 356). In December of 1856, Bessemer applied for a patent in America. Kelly, on hearing of Bessemer's application, challenged the U.S. Bessemer patent.

The U.S. Patent Office upheld Kelly's challenge a year later and awarded the U.S. patent to Kelly. The patent office made the decision based on Bessemer's own disclaimer in the application: "I do not claim injecting streams of air or steam into the molten iron for the purpose of refining iron, that being a process known and used before."[9] This amazing disclaimer in the U.S. application does not appear in his British application, in which Bessemer claims to be the originator. It may well be this was due to a unique difference in the patent applications of the two countries. In the United States at the time, it was required that you take an oath that you are the originator of the process. Taking an oath was a serious ethical and moral act and one of honor to the Victorian mind. In fairness, great mega-inventors such as Bessemer were more innovators than originators.

Historian John Boucher was able to clearly demonstrate this in 1923, years after Kelly's legal victory. In Boucher's book, *William Kelly: A True History of the So-called Bessemer Process,* Boucher concludes, "Let the future erase the name of Henry Bessemer from the cold and lifeless marble of the past, and carve in its stead, the name of William Kelly." The story, like most of the great Victorian inventions, is somewhat complicated, and representative of the innovative streams of the period. The Victorian era allowed for a free movement of information through scientific organizations and exhibitions. Kelly, in particular, gave many demonstrations in Kentucky to steel makers from the world over. The chemical basis for the process was that of Kelly. Kelly was a chemist and well understood the oxidation of carbon and impurities in iron. Kelly's insight, however, was more likely from his many years as a charcoal blast furnace operator. Pennsylvania furnace operators had in the early 1800s noted that cold air blasts could change the nature and properties of furnace iron. Excessive blasts produced what was known as "malleable iron" because it was softer and less brittle, probably even approaching steel in some cases.[10] Interestingly, Bessemer noted the same information in his early studies of British iron makers. Bessemer and Kelly were well aware of Neilson's hot air blast in Scotland as early as 1820. Still, Kelly would appear to be several years ahead on actually performing related experiments and demonstrating the pneumatic process in Kentucky.

Kelly's process was known in most metallurgical circles worldwide by the 1850s. This information allowed Bessemer to advance from behind in his understanding of the chemistry of the pneumatic steel making process. Steel historian John Casson, asserted, "'Bessemer was quite capable of originating the idea himself, but it would be strange if in eight years he had not heard something of Kelly's pneumatic process.' It was well known in Cincinnati, and letters passed between the iron men of Cincinnati and England every week."[11] Finally it should be noted that several historians note the disappearance after a year of two mysterious Englishmen in Kelly's employment.[12] The information flow of the Victorian era might have been slower than our Internet age, but its network was better focused in specific areas and not diluted by overflow.

William Kelly — Pittsburgh Irish steel maker and co-inventor of the pneumatic steel making process

The streams of innovations that led to the pneumatic process predate even Bessemer and Kelly. In the 1840s, two Englishmen, A. Perkins and R. Plant, had separate patents on the use of blowing steam to oxidize carbon out of molten iron. The Victorian amateur engineer and great Scottish painter and mega-inventor James Nasmyth approached the honor of originator of the Bessemer Process as well. In 1854, James Nasmyth improved on this process further by bottom-blowing molten iron in a converter to produce steel. Nasmyth's work had been discussed and reviewed at the meetings of the British Iron Association. Nasmyth went on to be credited as the inventor of the steam hammer as well as the father of Scottish landscape art. Nasmyth's drawings of his bottom blowing steam process would have greatly helped Bessemer, and it is clear Bessemer reviewed them in detail. All of these inventions had serious mechanical flaws that limited their use. Still the direction and focus on pneumatic steel making had started as early as 1820 by James Nielson.

Bessemer was a frequent visitor to the patent office, international exhibitions and manufacturing association meetings. It is known that Bessemer was well aware of Nasmyth's work, having read the patent and actually experimented with it.[13] Kelly, while isolated in Kentucky, was still part of the Victorian information network and would have also been aware of the work of Nasmyth. The very success of Henry Bessemer is a tribute to the Victorian scientific and engineering networks. Bessemer was able to assimilate the information and develop the necessary theory for his mechanical design without the experimental pain of Kelly's many experiments.

One of the best networks available to Bessemer was the British Association for the Advancement of Science (BAAS), which held regional meetings open to all. We know for a fact that iron-making technology was followed by the British Association for the Advancement of Science as early as 1841. The president of BAAS opened the annual meeting of 1841 proudly: "We have had experiments carried on at furnaces and iron-works, on railroads and canals, in mines and harbours, with steam-engines and steam vessels, upon a scale which no Institution, however great, could hope to reach."[14] This introduction was a reference to the steam-blowing process of Nasmyth to be discussed at the meetings. These discussions took place years before Kelly's experiments in Kentucky. In the late 1840s the use of Neilson's hot blast malleable iron was being discussed at BAAS regional meetings as well. These processes certainly contained the elements needed for pneumatic steel making. Maybe the real founders of the Bessemer Process were Nasmyth and Neilson.

Since Kelly owned the U.S. patent and history supports him as the process originator, why is it known as the Bessemer Process? Part of the story is that Kelly did not go to England to challenge the British patent. Kelly was more interested in the national acclaim than the dollars. Europe adopted the pneumatic process with much speed. In 1870, the United States had only three pneumatic steel plants, while Europe had 58. Another consideration was the work of the brilliant steel mill engineer Alexander Holley. Holley in the 1850s had gone to England to study the Bessemer experiments and obtained the licensing rights for the United States. Since Holley was the leading steel design engineer, he applied the Bessemer rights to his advantage in building the great pneumatic steel mills of Cambria, Troy (New York) and Pittsburgh in the 1870s railroad boom.

A third advantage was the dean of American steel, Andrew Carnegie. Carnegie, as a young railroad manager, had also gone to England to meet Bessemer. Bessemer demonstrated to Carnegie the superiority of Bessemer rails. In 1875, when Carnegie was to launch his great steel empire to supply the railroads, he commissioned Holley to build the largest steel mill ever at Braddock, Pennsylvania. Carnegie hoped also to avoid any legal battles by the use of the Kelly patent. This would forever ensure Bessemer's legacy. Near the end of Kelly's life in 1887, Carnegie visited the retired inventor to thank him for the process that made Carnegie a millionaire.

The real story is that Kelly's chemistry and metallurgy needed Bessemer's equipment to premier the steel-making process in the world. Kelly was an expert metallurgist, chemist and practical steel-maker. Bessemer was a mechanical genius and inventor who ultimately was awarded 120 patents. The creative streams that would lead to pneumatic steel making were already in place; even without Kelly or Bessemer, pneumatic steel making was only a few years away. Bessemer had earlier invented a stamp-cancellation machine for the British government, but contrary to what many biographers have written, he earned little from it. In fact, because he failed to apply quickly for a patent, Bessemer lost all commercial advantage. It was a lesson not lost on Bessemer in his quick application for pneumatic steel making. The stamp machine did allow him to be knighted. The Royal Society awarded Bessemer a medal for his invention of a sugar cane press, not his steel making process. What earned him a fortune, however, was his patent for the making of gold gilt paint. Gold painting was all in the fashion. History has awarded Bessemer credit for the steel-making revolution. Bessemer's interest in steel centered on the need for steel cannons, which were the fear of Europe. The Krupp success at the Great Exhibition of 1851 had stirred Europe. The pneumatic experimental processes lacked the mechanical means to go into production, and it was there that Bessemer made his mark.

Most historical accounts suggest that the Bessemer converter had a mechanical advantage of more rigorous stir than Kelly's system. Bessemer's equipment advantage was clear even though his understanding of the process was questionable. The licensing of equipment ultimately gave Bessemer the edge in the marketplace as well as in court. Neither Kelly nor Bessemer, however, produced workable steel on a consistent basis! The pragmatic process of pneumatic steelmaking required the genius of yet another Victorian metallurgist — Robert Mushet.

The amount of carbon removed by the Bessemer process depended on the length of the air "blow." This air blow was from 15 to 20 minutes. The converter operator watched the amount and nature of the sparks to estimate the amount of carbon removed. With enough time and air, the carbon could be lowered to the level of wrought iron. For good rail steel, .2 percent carbon composition produced the right amount of toughness, strength and wear resistance. This last point presented a minor technical problem, for even the Bessemer Process had variation. Carbon controls in steel making, as well as oxygen control, were major problems. Overblowing would reduce the carbon below the optimum .2 percent carbon, and that in turn would reduce wear and strength. In addition, overblowing increased the oxygen content in the steel. This oxygen in the steel, if not removed, made the steel brittle. Bessemer had learned this early on as he produced tons of bad steel.

The answer to the oxygen problem came in the 1857 by Robert Mushet. Mushet added a type of manganese ore (spiegeleisen) to the molten iron and solved the problem. The manganese cleaned the steel of the oxygen. Actually, this approach was used earlier in Prussia to produce high-quality steel. While Mushet was looking to control carbon and oxygen in steel, serendipity played a role in improved steel properties. Future metallurgists were to find that manganese strengthened and toughened steel in its own right. In addition, it helped tie up the tramp element — sulfur. The fact is, without Mushet's invention, neither Kelly nor Bessemer could have produced quality rail steel. About 15 minutes to 20 minutes of blow was needed to convert 10 tons of pig iron into Bessemer steel with the Mushet invention.

There are some other metallurgical fine points that caused poor Bessemer steel for rails. One of these led to the failure of the Freedom Iron and Steel Works of Lewistown, Pennsylvania. This plant, unlike the plants of Johnstown, Pittsburgh and Detroit, used eastern iron ore, which is naturally high in phosphorus. Phosphorus is a trace element that can embrittle steel. Worse yet, the Bessemer Process is incapable of removing it. The Bessemer converter could not control phosphorous at the time. The amount of phosphorus in steel was a result of the ore and coke phosphorus content going into the furnace. Freedom Iron failed in 1868 but was dismantled and reassembled in Joliet, Illinois. The rebuilt furnace worked successfully then with low-phosphorus ore from the Great Lakes region, which was also being used at Edgar Thomson Works. This accidental occurrence linked the problem to phosphorous in the ore and would play an important role in the spread of Bessemer steel making in Europe. Through all these industry failures and failed product trials, Edgar Thomson, president of the Pennsylvania Railroad, kept a vision of the potential of Bessemer steel rails.

Cambria Iron Works did not erect its first Bessemer converter until

1869, and the first Bessemer rails were rolled in 1871. Cambria did, however, hire Kelly to experiment with pneumatic steel making at the plant in 1858. It was at Cambria that Kelly became famous in the steel business and among the immigrant Irish. Kelly, as we shall see, was in every way the father of Braddock and Carnegie's future Edgar Thomson Steel Works. The Bessemer revolution would have been much delayed if left in the hands of Kelly, Bessemer, or Mushet. It would be the Victorian industrial dreams of Andrew Carnegie that spurred the Bessemer revolution.

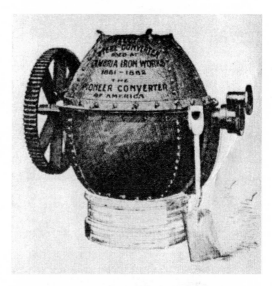

The Kelly steel converter used to pioneer pneumatic steel making at Cambria Iron Works

Railroad rails had evolved from wood to cast iron to wrought iron to steel. Cast iron offered great wear resistance because of its high carbon content. Cast iron was strong and could handle the load but was brittle to mechanical shock. Wrought iron, with its extremely low carbon content, could handle the shock of heavy, fast-moving trains. Wrought iron, which is as close to pure iron as it gets, was a bit too soft, however, for heavy railroad traffic. This softness could cause "snaking," which led to many wrecks. Snaking was caused by the pressure of the train curling the rail up, often into the train cars. The very fear of snaking reduced the speed at which trains traveled. Wrought iron also wore faster than cast iron. In the 1850s steel was an unknown engineering material. Steel had been made only by forging iron into swords and tools, as well as some small amounts in the crucible process. However, the crucible steel process was incapable of producing the tonnage needed for the railroads. Steel, like anything new, generated many concerns of the railroaders. Pittsburgh and Braddock were lucky to have a pioneering spirit in railroad executive Edgar Thomson, a chemical genius in Kelly, and a great salesman in Andrew Carnegie. Neither Thomson nor Carnegie was a technical man or engineer. Still, each was a true businessman. Without the railroads there was no significant market for steel except maybe in bridge building, and without the Victorian railroad boom, there would have been little need for pneumatic steel. Kelly's pneumatic steel allowed the rails for the locomotive to carry 7,000 tons of carloads versus 250 tons for iron rails![15]

Initially, Andrew Carnegie showed no interest in the production of Bessemer steel. It violated a basic Carnegie rule of never pioneering; however, two partners of Carnegie were extremely interested in Bessemer production, those being Tom Carnegie (Andrew's younger brother) and William Coleman. Tom Carnegie was Coleman's son-in-law and Homewood neighbor. Coleman and Tom Carnegie drove from Homewood to downtown Pittsburgh together every day. Tom Carnegie was the dealmaker with his beloved personality. In addition, Tom had that Victorian fascination with science and chemistry. Kelly's steel-making process was the first to be based on chemistry, and that furthered the interest of Tom Carnegie in pneumatic steel making.

James Howard Bridge, Andrew Carnegie's longtime secretary, felt the concept of the future Edgar Thomson Works was worked out between Coleman and Tom Carnegie during these many drives downtown to work. William Coleman was an amazing Pittsburgh capitalist. Born in New York, he came to Pittsburgh as an apprentice bricklayer. In 1845, he started a forge and rolling mill to make wagon axles. In the 1860s using these profits, he successfully invested in coal mines in Irwin, Pennsylvania. He further invested in oil wells north of Pittsburgh. In 1863, Coleman opened an iron rail rolling mill in Sharon, Pennsylvania. Besides these iron and rolling mills, Coleman was investing in railroads, such as the Allegheny Valley railroad. Coleman visited Europe in 1867 to study coke, rail rolling and the Bessemer Process. Upon returning from Europe, he made a trip to the Great Lakes iron ore mines. It is clear that no man in Pittsburgh better understood the railroad rail business and the Bessemer Process. Historian and personal secretary to Carnegie, James Howard Bridge points to William Coleman as the visionary of Edgar Thomson Works. The group of Pittsburgh businessmen interested in the Bessemer Process continued to grow.

Henry Philips was a boyhood friend of Andrew Carnegie and an amateur chemist. Philips's technical background made him key in eventually winning over Carnegie. Philips, as a partner in Carnegie's iron-making industries, was the first to employ a professional chemist in the manufacture of iron. Philips followed all the advances in chemistry, including the Kentucky experiments of Kelly. William Coleman also had spent years studying the possibilities of Bessemer steel and was eager to invest in the pneumatic steel process. Carnegie had made much from these friendships in the past, but he was also slow to get into the Bessemer steel business. Carnegie liked to research and listen to many different views. He was a very conservative investor. He hated to pioneer but wanted to be what marketing people call an early adopter. The Pittsburgh group around him was starting to remove his doubts.

The story of how Carnegie came to build the great Edgar Thomson

Bessemer steel plant was typical of his luck. In England, by 1870 the use of Bessemer steel for rails was booming. Edgar Thomson, president of the Pennsylvania Railroad, sent Carnegie to Europe to sell bonds. Of course, while there he could also look into and study the Bessemer steel industry. And he did exactly that, launching a major study of Bessemer steel making. Carnegie actually visited Henry Bessemer at his Sheffield Steelworks in 1872. Bessemer had been very successful in England with more than 20 Bessemer steel plants in operation at the time. Carnegie could see that Bessemer's steel investment was changing Europe, where railroads were rapidly switching to Bessemer steel. The salesmanship of the then fifty-nine-year-old Bessemer impressed Carnegie. One of Bessemer's favorite sales tools was an old cannon cast out of his steel for Krupp of Germany. That cannon had demonstrated the strength of his new steel. Still, Germany's great Krupp Works had rejected the Bessemer process, favoring the new "open-hearth" process. Carnegie needed an explanation for why Bessemer's steel process had not succeeded in the great steel mills of Germany.

The secret lay in Germany's high-phosphorus ores. Bessemer was an inventor and innovator. It was his mechanical skills, not his scientific knowledge, that had gained him success. Like Kelly, he had faced many early failures. In particular, Bessemer's early experiments produced very brittle steel. Analysis of the steel showed a high phosphorous content, which caused brittle steel. This started him on a sidetrack of trying to remove the phosphorous in the furnace, although he did not understand the source of the phosphorous. Bessemer did not realize that the source was from the English ores used to make the pig iron for the Bessemer furnace. The break that commercialized the Bessemer Process came accidentally when Swedish iron ore, which is extremely low in phosphorus, came on the market. These low-phosphorous ores made the resulting steel very ductile. This natural resource advantage of Swedish ores gave Bessemer a major commercial success, although he did not understand why. The Germans did not have a source of low-phosphorus ores. Trials in Bessemer steel making ended in disaster throughout Germany. Again, Carnegie's oldest partner, providence, joined him, for the new Great Lakes ores being used in Pittsburgh were extremely low in phosphorus, allowing for successful operation. Carnegie felt he could finally move ahead with the process that he would bring into domination.

While Carnegie's eventual acceptance of the Bessemer Process ensured its place in history, its use in central Europe required yet another metallurgical invention. German iron ores in particular were said to be so high in phosphorus that "they glowed in the dark." Alfred Krupp purchased the rights to the Bessemer Process but failed to make it profitable because of the high phosphorous. The technical problem became a major public-relations problem, crippling his ability to sell German Bessemer steel. Things got worse, as many of the great Krupp cannons failed in service because of high phosphorus embrittling the steel. Krupp's chemists and metallurgists had, however, found the secret — eliminate high-phosphorous ore. Part of the secret came from the American experience of Bessemer engineers with Freedom Iron's move to low-phosphorus western ore. Krupp secretly started to buy up phosphorus-free Spanish mines. He also secretly constructed a Bessemer steel plant under the code name C & T Steel. Through this clandestine operation, Krupp used the Bessemer Process to supply an unknowing Europe its ordnance.

The solution for Germany came in 1876 with the work of two British metallurgists, Sidney Gilchrist Thomas and his cousin Percy Gilchrist. Even Britain was concerned because its sources of phosphorus-free ore were being used up rapidly. The answer was in the use of "basic," or limestone, brick in the converter lining. This created a basic slag on the molten metal that removed phosphorous. This became known as the Gilchrist-Thomas Process. Without this process improvement, it's likely that Bessemer steel would have died a commercial death throughout Europe and even in England. One European historian said, "For Europe as a whole Thomas's discovery had an importance comparable with those of Bessemer and Siemens. When Bessemer himself died in 1898 he was an immensely rich and renowned figure, but the world's production of steel by his original process had begun to decline in face of competition from the more widely applicable Gilchrist-Thomas process."[16] This is typical of the evolution and adaptations of inventions during the Metallurgic Age.

With Bessemer steel gaining railroad orders throughout the world, the proposition of building a Bessemer steel rail plant now became a necessity for Carnegie. Carnegie had seen the huge profits of the English steel-making operations. He had talked to Bessemer himself and had proved the process. He understood the problem of Krupp with German ores. Carnegie had also seen the outstanding performance of the rails on the roads. More important, Carnegie wanted Bessemer's proven equipment patents. All of Carnegie's friends had become strong supporters of the pneumatic steel-making process. In addition, building a Bessemer steel plant would create a huge customer for Carnegie's pig-iron production at Lucy Furnace. The newly converted Carnegie returned from Europe determined to build a plant. While in Europe, Carnegie's brother and friends purchased the site of the defeat of General Braddock and a young George Washington during the French and Indian War for a Bessemer mill. Andrew Carnegie was 38 years old at the time of the purchase of Braddock's field. It was Carnegie who required that the new plant be named after his old friend, mentor and future customer Edgar Thomson. With the finances set, it was Carnegie's vision that would make Edgar Thomson Works famous. Carnegie did every thing in a world-class manner. While other steel makers built steel mills to be competitive and efficient, Carnegie's mission was to be the best in the world. Braddock would bring pneumatic steel to the pinnacle of Victorian manufacture. Carnegie's Edgar Thomson Works at Braddock produced its first heat of steel in 1875 and continues to produce steel to this day.

Braddock needed more than the process of Kelly and Bessemer, the rail-roads of Edgar Thomson, and the finances of Carnegie — it needed engineering genius. Steel had to take over the wrought-iron steel market in rails. To that end, Carnegie hired the best Bessemer engineer in the world, Alexander Holley. Holley in 1862 had visited the Bessemer steelworks at Sheffield. Holley was an exceptional metallurgist and engineer and a prolific technical writer who had built most of the operating Bessemer mills in this country. He was the American agent for the use of the Bessemer Process and equipment used, in particular, in two of the best plants in the world at Troy and Cambria. Other than possibly Henry Bessemer or William Kelly, there was no better person to build Carnegie's dream. In typical Carnegie fashion, Holley was given a free hand to build the best operating mill possible. The plant was to have two six-ton converters capable of producing 75,000 tons of Bessemer steel per year. Carnegie manned his new plant with Cambria Steel managers trained by William Kelly. One of those Kelly men was Bill Jones, the future manager of Carnegie's Edgar Thomson Works. Jones would do for pneumatic steel-making what Henry Ford did for the assembly line.

The future of Braddock's Edgar Thomson Works would determine the future of the American steel industry. No place on earth would produce as much Bessemer steel as Braddock, Pennsylvania. In manufacturing heavy-gauge and light-gauge rails from 1875 to 1913, Edgar Thomson Works would produce enough rails to lay track to the moon and back several times! At its peak from 1890 to 1910, Edgar Thomson Works produced 25 percent of all the steel made in the United States. By 2003, it could be said that no plant on the face of the earth had produced as much steel as Edgar Thomson Works. Today, Edgar Thomson Works has the last operating blast furnace in the once-great Monongahela Steel Valley.

Neither Pittsburgh (Kelly's hometown) nor Braddock, the city he built, has plaques to honor Kelly. While few plaques in Detroit and Johnstown remember Kelly, only a handful of Irish steel-makers honor the true history of the "so-called Bessemer Process." The town of Braddock does stand as Kelly's greatest monument. Kelly retired to Louisville, Kentucky, after losing the patent fight in 1871. Near the end of his life in 1888, Carnegie came to visit and thank Kelly, the man who had made him millions. In World War II, Carnegie's Monongahela Valley steel mills, centered around

Braddock's Edgar Thomson Works, would outproduce the whole of the Third Reich. Having created most of America's industrial wealth and contributed to American victories in two world wars, the Victorian metallurgist William Kelly deserves a major place in history. Maybe the greatest tribute to Kelly is that his chemical process continued long after the death of the Bessemer Process in the 1950s. Today's huge chemically basic oxygen furnaces are based on Kelly metallurgical principles. More than 90 percent of all the steel ever produced in the world has been and continues to be made using Kelly's chemical processes as a foundation.

In 1870 there were 115,000 miles of railroad, with 30 percent being steel, but in 1900 there were 258,000 miles, and 93 percent were Bessemer steel. The penetration of steel rails in Europe had been even faster. Bessemer steel making allowed for the large tonnage production and lower costs not possible via the crucible process. While the growth in steel was enormous until 1890, wrought iron remained the engineering material of the Victorians. Steel production until 1890 could not meet demand, and its higher cost kept wrought iron and cast iron competitive.

The father of the steel age was neither Kelly nor Bessemer. It wasn't even Mushet or Thomas, who made the Bessemer Process work. It was a Braddock resident and plant manager of Andrew Carnegie's first steel plant, Edgar Thomson Works—Captain Bill Jones.

Jones, who lacks even a single plaque, propelled steel into the engineering material of the 20th century. Jones made Bessemer steel cheap enough to replace wrought iron in not only rails, but in tinplate as well. Bessemer steel owes as much to the economic development work of Jones as to Mushet's and Holley's technical developments. In 1876, Jones suggested the use of steel in tinplate production and worked with a nearby tin mill to produce the first steel tinplate. It was Jones who reduced the base cost of Bessemer steel below all ferrous competitors. Jones was a managerial genius and a premier inventor. In the 1880s, the steel-making flow at Braddock required large shipments of cold pigs of iron from eight miles away at Pittsburgh. The pigs were then remelted and charged as liquid (molten) iron into a Bessemer converter. The handling and melting added considerable cost to the steel produced. Jones invented a holding vessel for liquid iron. Blast furnaces could be tapped and the molten metal stored for use in the converter. This piece of equipment was known as the "Jones Mixer." Steel makers of the period hailed it as important as the Bessemer converter itself: "The importance of the [Jones] innovations could not be overstated, according to James Gayley, who was the superintendent of Edgar Thomson's furnaces under Jones and considered his boss's contribution to steel making on a par with Bessemer's."[17] Jones never saw the benefits of his invention, having died in a blast furnace accident in October of 1889.

Bessemer converters at Edgar Thomson Steelworks; 1885

A few days after Jones's death, Carnegie's right-hand man, Henry Clay Frick, and Carnegie's lawyer paid Jones's grieving and handicapped widow $35,000. Biographer Peter Krass said, "It was quite a bargain, considering that six years later Carnegie would estimate the Jones Mixer saved the company $150,000 to $200,000 per year, and that over the next decade, the company would spend hundreds of thousands of dollars in legal fees to protect its patents."[18] The Jones family tried some years later to obtain more money from Carnegie, but he considered it just another shrewd business deal.

In the end the names of Jones, Kelly, Thomas, Gilchrist and Mushet are long-forgotten. Bessemer is remembered in history books, city names, building names, street names and corporate names. Bessemer in history appears as a heroic inventor. The stream of invention and innovation, while forgotten, is key to understanding the advance of steel making. More important, it teaches us lessons of creativity. The pneumatic steel process and the economies of scale offered by mills such as Edgar Thomson lowered the price of steel, making it competitive with wrought iron. Earlier cast iron had been replaced by wrought iron, and by 1880 steel was king. As we have seen, the railroads supplied the initial stimulus, and the structural properties offered new markets in bridge building, construction, ordnance, tinplate, and shipbuilding. Bridges, in particular, were overtaken by steel. Steel offered a unique combination of strength and flexibility, being the intermediate product between cast iron and wrought iron. Steel has eight times the strength of wrought iron, and that offered the potential for long spans. The man that brought steel into bridge building was James Buchanan Eads, a wrought-iron capitalist.

Eads was already a famous and wealthy engineer prior to proposing the first steel bridge at St. Louis. Eads's submarine diving bells in the 1850s had spawned a very successful salvage business. The bridge design called for three steel arches. The use of steel offered a challenge equal to Eads since steel had not been used in bridge building. Even in the world's greatest steel-producing nation, England, steel was banned for use in bridges until 1877. Eads contracted the Keystone Bridge Company of Pittsburgh, but it also lacked experience in steel. Amazingly in 1870, Keystone had a young vice president that was very suspicious of steel–Andrew Carnegie! Carnegie, however, wanted the business, and actually went to Europe to sell bonds for the project. Carnegie also got Junius Morgan, the great financier, involved in the sale of bonds. Eads made many trips to Pittsburgh to ensure processing and testing met specifications. Carnegie had no steel mills in 1870 and had to purchase Bessemer steel from other suppliers. Carnegie's Keystone had numerous puddling operations to produce and roll wrought iron, it but had no equipment for steel. While Carnegie purchased the puddled steel, he had to build new rolling equipment.

Eads pioneered the use not only of steel, but also of chromium steel. Eads, however, was not the inventor of chromium steel. That honor goes to Michael Faraday. Faraday worked on a number of experiments between 1819 and 1824 using titanium, nickel and chromium. He studied combinations of steel alloys in an effort to reproduce the famous "Wootz" and Damascus steels of the ancients. With the aid of Sheffield crucible steel makers, he did produce some small quantities of alloy steel, including chrome to be used as razors and cutlery. Faraday published his experiments in many of the popular journals, which led to further experiments on chrome in steel by Mushet. It was, however, an American, Julius Baur, who patented chrome steel and produced it on a commercial scale, in 1865. Baur's work was done in a crucible furnace at his Chrome Steel Works in Brooklyn. Baur by 1869 was selling chrome steel for jail bars, burglarproof vaults, and mining tools. Eads followed developments in Europe and America, and he was confident that chrome steel would supply another factor or two of strength. Eads would work personally with the steel producers in Johnstown and New York to ensure its quality.

After the completion of the Eads Bridge in St. Louis in 1874, the next innovative use of steel was the Brooklyn Bridge. The Brooklyn Bridge was the first major bridge made of Bessemer steel. While the bridge had started during the construction of the Eads Bridge, it was not completed until 1883. The designer and chief engineer was Washington Roebling, the son of John A. Roebling. The Brooklyn Bridge would be the first to use steel wire cables. Washington and his father had already established themselves as experts in the building of wire suspension bridges using iron. They took these streams of development and expanded the market for steel, including the crucible process. The very special steel wire was produced in crucible furnaces and drawn into wire at Haigh Works in Brooklyn. The Roeblings went further by galvanizing the steel wire to prevent rusting. Galvanized steel is a much-overlooked technical advance of the Brooklyn Bridge. Passing the wire through molten zinc, which coats the wire, did the galvanizing. Galvanized steel has become the mainstay for protecting steel. Zinc coating is unique in that it not only acts as a coating barrier to oxygen and moisture but also chemically protects the steel even when scratched. Today 50 percent of all the world's zinc production goes to galvanizing steel.[19] Still the Brooklyn Bridge was far from a "steel bridge." It contained cast iron, wrought iron, and steel intermixed in its design. The 1890 Forth Bridge in England probably should be given the title of the first complete steel bridge. John Fowler, designer of the Forth Bridge, said had he used iron instead of steel, the Forth Bridge would have "both weighed and cost twice cost as much." In the end, steel dominated the world of Victorian railroads, bridges, and construction. The Brooklyn Bridge ushered in the Age of Steel.

6

Victorian Wars and Technology

The Civil War was a defining event of the Metallurgic Age. Technology vaulted forward in the steel, iron, coal, and railroad industries. In addition, the Victorian philosophy of bigger and better found application in the American killing fields. The American armies of 1860 started the war with Napoleonic armaments and strategies, but closed the war with the technology of the next century. The best summary of the world Victorian view of the Civil War can be seen in the fictional opening of Jules Verne's *From the Earth to the Moon* (1865):

> During the Civil War in the United States, a highly influential club was founded in the city of Baltimore, in the middle of Maryland. Everyone will remember the vigor with which that nation of ship owners, shopkeepers, and mechanics discovered their instinct for warfare. Simple businessmen leaped over their counters to become captains, colonels, generals, without ever having studied at West Point. In no time at all, they mastered "art of war" and, like their colleagues in the Old World, they won battles through lavish expenditures of bullets, millions, and men.
>
> Indeed, in one branch of military science — ballistics— the Americans even surpassed the Europeans. It was not that American weapons attained a higher degree of perfection, but rather that they were built bigger and so could shoot farther. In such matters as grazing, plunging, or direct fire; oblique, raking or flank fire, the English, French, and Prussians have nothing to learn; but their guns, their howitzers, their mortars are mere pocket pistols compared to the awesome mechanisms of the American artillery.
>
> This should not surprise us. The Yankees, the world's greatest mechanics, are engineers the way Italians are musicians and Germans are metaphysicians— by birth. It was only natural for Americans to take their bold ingenuity with them when they ventured into the realm of ballistics. And they developed their gigantic cannon, far less useful than their sewing machines, but equally amazing and much more admired. Witness the marvels of Parrott, Dahlgren, and Rodman. The Armstrongs, the Pallisers and Treuilles de Beaulieu could only bow to their rivals across the sea.

Verne's review gives a powerful testimony of the Victorian mind. (Note that Verne makes a political point by ignoring the Krupp.) The concept of bigger, better, faster, stronger, etc., is articulated perfectly by Verne. Victorians loved the power of technology, and perceived power in taking technology to new heights. The Civil War was a Victorian war that based success on the application of ever-new uses of technology. The European countries had hundreds of correspondents following the war because it was a huge technological experiment and proving ground.

To view the Civil War as a technological race, as many Victorians saw it, changes your perspective on some of the key players, such as Abraham Lincoln. Lincoln is remembered mostly for his spiritual, moral, and philosophical leadership. What is often missed is his love of technology and his boldness in applying it. One historian noted, "There was a streak in Abraham Lincoln that reflected young America's delight in the Machine Age, a streak that had more in it of Eli Whitney than of Henry Thoreau." It would be Lincoln more than any engineer or general that advanced the technology of war. He loved to test rifles, visit foundries, and attend artillery tests. He was a serious reader of the *Annual of Scientific Discovery* and *Scientific American.* Lincoln believed that victory would go to the side with a mechanical advantage. Yet his key generals tended to be extremely conservative in applying new technology, willing to fight a Napoleonic war. Lincoln did manage to develop a key friendship with Captain John Dahlgren, commander of the Washington Naval Yard. Dahlgren was a gifted designer of weapons, and he had a passion for new technology. Dahlgren would be Lincoln's advisor and friend throughout the war.

Lincoln knew that the South lacked the machine shops and tools to fight a technological war. While Lincoln's generals hesitated, the mechanics, engineers, and Yankee inventors pushed this advantage. John Ericsson (designer of the ironship the *Monitor*), for example, wrote the following to Lincoln:

> The time has come, Mr. President, when our cause will have to be sustained, not by numbers, but by superior weapons. By a proper application of mechanical devises alone will you be able with absolute certainty to destroy the enemies of the Union. Such is the inferiority of the Southern States in a mechanical point of view that it is susceptible of demonstration that, if you apply our mechanical resources to the fullest extent, you can destroy the enemy without enlisting another man.

Even more critical were the skills of the Yankee soldiers in innovating, adapting, and repairing equipment of all kinds. The Northern mechanical advantage was even seen in the most basic things, such as shoes. The North had the McKay Sewing Machine, which turned out tens of thousands of shoes. The South, lacking automation, had to exempt all cobblers from military service to try to keep the soldiers in shoes. The most amazing thing about the war was the ability through determination of the South to hold off the mechanized North for so long. They might have even defeated the North if Lincoln had not forced the North into applying technological advantage. In the early Napoleonic battles of the Civil War, the South prevailed.

In 1863, at a stream in Georgia known as Chickamauga, a Union regiment called Wilder's Lightning Brigade tore into the Confederates. The Confederate veterans of General Longstreet had never faced such a hail of bullets. Never before had the fight been so one-sided in favor of the Union. If President Lincoln would have had the total decision on arms selection, Chickamauga might have been remembered as the beginning of the end of the war. The carnage was so overwhelming that one colonel noted, "I had it in my heart to order the firing to cease to end the awful sight." The new rifle was the breech-loading, repeating Spencer. The Union soldiers had those guns because of President Lincoln, even though the Federal Ordnance Department opposed their use. In 1861, Lincoln and his secretary, William Stoddard had conducted their own tests on the White House's South Lawn. This area near today's Mall became known as the "White Lot" and was Lincoln's personal testing ground, but he shared it with herds of cattle needed to feed the troops. The tests of Lincoln and Stoddard that September morning in 1861 revolved around the technical question of muzzleloaders versus breech-loaders.

Muzzleloaders had been the rifle of the Revolutionary and Napoleonic wars. At the start of the Civil War, the Springfield muzzleloader was regulation-issue for soldiers. The muzzleloader musket was a known gun, and in many aspects a dependable rifle. It performed well in organized Napoleonic formations, which allowed time and organization for loading and firing. The Civil War was different; the tactics included massive charges against dug-in field fortifications. The morning that Lincoln and Stoddard ran their own tests, they achieved an average of three rounds per minute, but it required the experience of a frontiersman like Lincoln. For the young secretary it would be difficult to get a round a minute. Lincoln used a Henry repeating breech-loader that held 16 rounds and fired two a second. The muzzleloader required a set of steps to load the powder charge and ball down the barrel of the rifle. The packet had to be properly rammed or packed as well. The gun had to then be primed, cocked, aimed and fired. Mistakes were easy to make as Lincoln and Stoddard noted for themselves the ease of jamming the barrels of muzzleloaders. The problem could become monumental with raw recruits. Napoleonic armies spent years of training and drilling on muzzleloading. Lincoln was right with his precise observations on gun jamming in muzzleloaders, and those would be proved a few years later at Gettysburg. Historian Robert Bruce noted the results of a study of the Gettysburg battlefields: "If he made any of [a number of] mistakes, his gun was useless, probably for the rest of the battle. On the field of Gettysburg more than twenty-four thousand loaded muskets and rifles were found. Six thousand of them had one load apiece, twelve thousand had two loads each and six thousand had three to ten loads. One famous specimen had twenty-three loads rammed down in regular order."[1] This study clearly showed that young soldiers in the heat of this great battle had jammed their rifles, and maybe they even believed they were firing them! Breech-loaders held automatic cartridges to eliminate the drill of loading and packing. Jamming was also eliminated, which resulted in rapid fire.

The repeating breech-loaders gained their advantage based on rate of fire versus accuracy. During the Revolutionary War, the British, with well-trained and disciplined soldiers, achieved one hit per five musket balls fired. This is known as the hit rate or "shots per casuality." This rate declined to 1 for 20 during the Napoleonic wars because of the lack of training, and the need for conscripts. In the Civil War the casuality ratio continued to decline with the decline in training. It is suggested that in some battles it was one casuality per 200 shots.[2] Success with muzzleloaders depended on en masse Napoleonic formations concentrating firepower. The repeating rifle significantly upped the firepower, and thus the casualities. Furthermore, the repeater did not require battlefield stands and drills for reloading. The army was therefore more mobile.

The Federal Ordnance Department's opposition to repeating breech-loaders, which would eliminate the problem of drilling, centered on several points. First, in 1861 there was little manufacturing capability for producing any sizable orders of breech-loaders. The tooling required for production would require a major redesign of factories. This limited capacity, while it worked against the Lincoln plan to convert to breech-loaders, probably saved the Union. Jefferson Davis and the confederacy were early believers in breech-loaders. The Confederacy suffered more from the lack of tooling and basic industrial assets to have the ability to make the conversion. Had the Confederates been fully able to make the conversion to breech-loaders by 1862, there would have been no Gettysburg Address! The Union, with the exception of Lincoln, opposed breech-loaders as too novel and difficult to produce. Furthermore, the Union generals continued to argue that such rapid fire from repeating breech-loaders could lead to mechanical jamming, and also that such a complex mechanism could not be corrected in the field. The cost of a breech-loader was thirty dollars versus nineteen for a muzzleloader. Union military experts even used the excuse that soldiers would have to carry more ammunition, which would slow their march. Lincoln had to be amused at this one since the Union troops had hardly moved in two years. The arguments are familiar to any manager or inventor that proposes technological change. Lincoln, however, was not deterred in his quest to arm with breech-loaders. He wrote orders over the generals, searching for manufacturing capacity himself. Luckily, he thankfully found Christopher Spencer.

Christopher Spencer is rumored to have been the model for the hero in Mark Twain's *A Connecticut Yankee in King Arthur's Court,* and with good reason. Spencer had as a young boy improved on his grandfather's muskets, making them very accurate. At 15, using information from his grandfather's volume of *Comstock's Philosophy,* he built a steam engine. In his late 20s he had built a steam automobile and was driving to and from work at the Colt gun factory in Hartford. Spencer was even experimenting with flying apparatus in the 1850s. Spencer embodied the heroic image of the Victorian engineer and meta-inventor, a dreamer and machinist who could build anything his imagination could envision. Like Bessemer and Nasmyth, Spencer had the cognitive ability to tap into a stream of creativity and invent. He was a perfect model for any of Jules Verne's characters as well. Working as a machinist in the Colt gun factory, Spencer honed his invention of the repeating rifle. The Spencer repeater used a metallic tube to hold seven rounds. The seven rounds could be fired in two to ten seconds, and then pulling a lever would load another cartridge of seven rounds through the stock. The only limitation noted in military tests was the overheating of the barrel, but a sustained rate of 14 rounds per minute created no problems. Lincoln was sure of the role that a repeater could play and was confident in its safe use from his own trials with the repeater. Lincoln confirmed this safe rate of fire at his "White Lot" testing ground, and even suggested a better gun sight for the Spencer repeater.

The military continued to drag its feet and stalled the order. Finally, at the end of 1861, Lincoln, by executive order, purchased 10,000 Spencer repeaters and another 25,000 Marsh breech-loaders. Spencer needed months to get the needed machine tools for manufacture because many had to be ordered from England. When the Union army got the repeaters in 1863, it changed the battlefield. *Scientific American* hailed the Spencer rifle, and wrote often on its success. Early tactics of the Federals were extremely deadly. They could fire one volley and wait, fooling the Confederates into believing they were reloading. As the Confederates charged, the Union soldiers would unleash another six rounds. General Custer developed a number of infantry and cavalry tactics to fire the repeaters on the move. Not only were the Confederates stunned, but so were the foreign correspondents. The era of the muzzleloader had ended because of Abraham Lincoln. Actually, Lincoln had gone further by becoming the first to purchase a battery of machine guns.

Lincoln again ran personal tests on these new machine guns, known then as "coffee mill guns." Like all these guns of ingenuity, there were a number of inventors. J. D. Mills proved to be the best salesman of the group. Mills had also won the endorsement of two American engineering giants— Peter Cooper and Cyrus Field. These guns could deliever 50 rounds per minute. The generals stalled and avoided Lincoln's planned trials. Finally Lincoln personally signed for a purchase from Mills of 10 guns that Mills had in inventory. Lincoln ordered 50 more by going around the conservative War Department. The guns saw limited action because the military continued to sabotage their use. Again we see that had Lincoln been the only decision maker the war might have ended much earlier. The more famous Gatling gun was not brought in till the very end of the war, but the "coffee mill guns" of Mills had inspired it. Dr. Richard Gatling patented his design last in 1862. The Mills and Gatling guns had little differences, but the Gatling gun seemed to be of higher manufacturing quality. By the end of the war, Gatling's gun had won over the military bureaucracy, but too late for much service. Still, it did see action at Petersburg. A year after the end of the war, the Gatling gun won official recognition for the American army. Lincoln, however, deserves the credit for its adoption. The French used the Gatling gun successfully during the Franco-Prussian War, and it found wide use by the Americans in the Spanish-American War. Had Lincoln headed the War Department with a free hand, he might well have shortened the Civil War by years with visionary technology.

While the rifle situation at the beginning of the war was stuck in the technology of the previous century, artillery was a nightmare of ancient and modern technology. Lack of standardization was initially a bigger problem than the technology. The combinations of guns, permutations of gun types, and variation of ammunitions inhibited the army's ability to fire anything. Gun barrels were made of bronze and iron, with a few advanced steel ones. Barrel designs could be smoothbore or rifled in any of the three materials. There were at least 11 oveball manufacturing designs, from Napoleons to Parrotts. These manufacturing designs could come in at least nine calibers and seven different shells. The confusion over models, logistics, and lack of coordination crippled artillery units early on. While the conservative head of Union Ordnance, General Ripley, had frustrated Lincoln's new rifle technology, his desire for standardization brought order to the artillery situation. As if this weren't enough, the same breech-loading versus muzzleloading debate was under way with larger artillery.

At the start of the war, the Napoleon 12-pounder field artillery accounted for almost half of all the cannons on both sides. It was a cast-bronze, smoothbore field gun developed in France in the 1850s. It was actually named after Napoleon III, and had range of 1,619 yards for a 12-pound ball. The Napoleon was considered standard artillery throughout the world.

The Napoleon's main competition came from the 10-pounder, 3-inch-diameter wrought-iron rifle. To form the 10-pounder, broiler plate was wrapped around a bar and welded. Both were considered light smoothbore artillery. General Lee in particular favored the Napoleon, but when the Federals took possession of the Tennessee copper mines (bronze is 90 percent copper and 10 percent tin), the South stopped production. Also, at the start of the war, Mexican War ordnance such as the 6-pounder bronze cannon was extremely common. Bronze guns in general were expensive compared with cast iron and wrought iron. Bronze lacked the strength of wrought iron, which limited its range. The ability of a field cannon to project more weight farther depended on the strength of the breech in handling greater powder charges. Stronger designs and materials became the quest of artillery engineers and the popular press.

The first stronger breech design appeared in the 1850s. Union engineer Robert P. Parrott developed it. Parrott operated a foundry at Cold Spring, New York. His new cannon was based on a cast-iron rifled barrel with a strengthened breech, which had a wrought-iron band around the beech to strengthen it. This band or loop was welded and hammered in place. These guns, known as Parrotts, came in 10 — and 20-pounder versions with a range of more than 2,000 yards. Parrott patented the gun in 1861, but the Confederacy quickly copied it in Virginia foundries. One of the first to see the new gun at the foundry was Abraham Lincoln, who for once had the support of his generals in the purchase of Parrotts. The larger Parrotts had a tendency to burst at the end of the reinforced band, requiring repair welding if overcharged. However, another major advantage of the Parrott was rifling. In the late 1840s, the Krupp Works of Germany had mastered rifling of the cannon bore. Rifling put spiral grooves in the barrel. Rifling of cannons, as in small arms, gives spin to the projectile versus the tumbling that occurs in a smooth bore. This spin allows accuracy and improves distance.

The search for bigger cannons was a natural one for the Victorian era, and had started prior to the war. The Victorian name for such huge cannons was "Columbiads." The work of Naval officer John A. Dahlgren deserves special comment. In the 1850s, Dahlgren invented a bulbous breech using cast iron. It was nicknamed the "soda water bottle." The design allowed for massive powder charges because of the breech design. These 100-pounders made Dahlgren famous, and allowed the Union to start the war with very powerful siege Columbiads. Still, it was another military genius, Major Thomas Rodman, who would cast the greatest Columbiads ever. When Jules Verne looked for science to back his fictional cannons to shoot a projectile to the moon, he proposed a massive Rodman cannon. The following is a discussion from Verne's novel *From the Earth to the Moon*:

"Easily, since I was on the Experimental Committee during the war," the general reminded them. "I should point out that Dahlgren's hundred-pounders, which had a range of 2,500 fathoms, gave their projectile a muzzle velocity of 500 yards per second."

"Good. And Rodman's Columbiad?" the president asked.

"In its initial trials at Fort Hamilton, New York, the Rodman Columbiad shot a projectile weighing half a ton a distance of six miles, with a muzzle velocity of 800 yards per second. Such results where never obtained by Armstrong and Palliser in England."

"Oh! The English!" exclaimed J. T. Maston, shaking his formidable hook at the eastern horizon.

"So," said Barbicane, "800 yards must be the greatest velocity reached so far? ... So let us take this muzzle velocity of 800 yards per second as our starting point. We must multiply it twenty fold."

Verne's summary of the technology is not only accurate, but also shows how the Civil War technology had captured the imagination of the world. The large Columbiads exemplified the quest of the Victorians for bigger and better. Verne's idea of a cannon capable of firing a projectile to the moon was typical of Victorian dreams.

Rodman had developed his Columbiad in the late 1840s. He used a streamlined bottle design with a large breech diameter. The revolutionary part of the Rodman Columbiad was not in the design, but in the processing. The processing was a radical break from normal cannon production, so much so that the South's largest foundry, Tredegar Works, refused to manufacture the design because of process-control concerns. Rodman's process pioneered the concept of cooling his cast-iron cannon from the inside out instead of vice versa. Such cooling caused layers of cast-iron metal to shrink like banding, thus strengthening the metal. Rodman casted his hollow cannon with a removable center core. (Jules Verne used this same method in his moon gun.) The core was cooled using water or air while the cannon remains in the sand mold. The method slowed down the production of the mammoth Columbriad. Jules Verne calculated it would take weeks for his fictional Columbiad to cool. In *From the Earth to the Moon*, he notes, "Now when Major Rodman had cast his 160,000-pound cannon, it actually took fifteen days to cool. How long, then, would this new monstrous Columbiad, crowned with clouds of vapor and defended by its terrific heat, remain hidden from its admirers' gaze?"

Rodman's early guns had 8-inch and 10-inch bores. The 10-inch Rodman could fire a 128-pound projectile 1,800 yards. They were the first guns to see action in the defense of Fort Sumter in 1861. The Rodman would prove to be a consistent gun, not suffering the breech explosions of the large Parrotts. The early success of the Rodman design spurred even bigger dreams. In 1861, Rodman cast the largest cannon in existence, a 15-inch

Columbiad. The 15-inch gun could fire a 428-pound projectile 4,680 yards. Ultimately Rodman developed a 20-inch gun capable of jetting a projectile four miles. The size pushed the efficiency of gunpowder at the time. The first 15-inch Rodman was put in defense of Washington, DC in an Alexandria battery containing five additional 200-pounder Parrotts. At the end of the war, Rodman produced two even larger 20-inch cannons. These huge Columbiads never saw active service but remained at military readiness into the 20th century. The standard Rodmans (8 and 10 inch) were ordered in large quantities.

Verne's moon cannon–casting scene was based on the casting of the 20-inch Rodman at the Fort Pitt Foundry at Pittsburgh. The Fort Pitt Foundry had only one rival in the production of massive guns: the Trageror Foundry in Virginia. The Fort Pitt Foundry produced 60 percent of the Union's large guns, and in the 1860s it overtook Krupp in Germany for the casting of the world's largest guns. The Fort Pitt Foundry castled some of the most famous guns of the war, including the 13-inch mortar called the "Dictator," which hurled 218-pound balls two and half miles during the siege of Petersburg. Reporters from *Scientific American* and *Harper's Weekly* made almost weekly reports of the activity at the foundry. The foundry drew visitors from all the foreign capitals who wanted to see the behemoth in operation. One author described the interest:

> There is probably no single establishment in the United States that attracted so much public attention during the war as the Foundry. It is thronged daily with visitors. Many traveling strangers in passing would delay their journey a day or two in order to visit the works. Distinguished military and naval officers from England, France, Spain, Russia, Sweden, Denmark, Prussia, Sardinia, and Austria, who had come from Europe to observe the operations of our armies in the field, or to note the progress of the war and the manner of conducting it, came from Washington City for the special purpose of examining the works, and of witnessing the casting of monster cannon.

The mammoth 20-inch Rodman cannon cast at Fort Pitt Foundry in 1863 (*Scientific American*)

The Fort Pitt Foundry captured the focus of the world so that writers like Verne in France had no trouble following its manufacturing accomplishments.

The casting of the giant 20-inch Rodman gained the attention of the world in February of 1864. The foundry got its pig iron from the Hanging Rock region of Ohio and Juniata region of Pennsylvania, which were the nation's primary charcoal furnace districts. It then remelted the cold pigs in two cupolas and poured them into six holding furnaces. The 100,000-pound weight of the great gun would require the sequential pouring of the six furnaces, which Krupp Works in Germany had pioneered and Verne reported in his fictional novel. The foundry was overcrowded with reporters noting every detail. The final piece weighed 116,497 pounds and cost $32,718. The gun today is preserved at Fort Hamilton in Brooklyn, New York. The foundry did produce more Rodmans and a series of very powerful Dahlgrens for the Navy. These great Dahlgrens were also 20-inchers, but could hold a slightly larger powder charge. The guns were named Beelzebub, Satan, and Lucifer. With the Civil War over, the guns were sold to Chile for its war with Peru. These guns represented the height of Civil War technology.

Lincoln took technology to the cutting edge, and with that came some failures as well. Lincoln, however, never hesitated to allow trials of military gadgets. One of these was the steam cannon. The idea was to use steam power to propel a projectile. Actually, a Yankee inventor, Jacob Perkins, had invented the steam cannon in the 1850s. The steam guns were a failure, but Lincoln at least got trials. In fact, Lincoln actually took part in one experiment on the Perkins cannon. Lincoln cranked the device; the balls were fired but bounced off the target, bringing Lincoln and the crowd to laughter. There remains an example of the steam cannon on Hilton Head Island. Another unusual cannon was the double cannon. The double cannon was to simultaneously fire two balls connected by a chain. The idea was to hurdle a twisting-chained double shot. Trials of the double cannon failed for both the North and South. One of these double cannons can be found today at Athens, Georgia. The devices peddled to Lincoln seemed endless, including tanks and body armor. There were a few areas, however, where Lincoln could not take technology to the cutting edge. Two of these were steel cannons and breech-loading cannons. Both of these technologies were the intellectual property of Alfred Krupp. Krupp studied the Civil War with a number of agents, which helped him to become the armor king. Krupp's guns crushed the French in the 1870s, but the French learned from the beating.

The French company Schneider emerged as Krupp's major competition. These two great companies would compete for the arming of the world

in the last decades of the Victorian era. Jules Verne had envisioned this great rivalry in his book, *The Begum's Fortune* (1879). Even more amazing was Verne's prediction of the long-range gun, which had started with the development of Rodman's 20-inch gun and ended with "Long Tom" of the Boer War (1899–1902). Verne wrote of long-range shells from miles away whistling into a fictional town, Franksville. Verne's story echoes the real-life battle of Ladysmith during the Boer War. The Boer War had several future writers as observers, such as Arthur Conan Doyle, Rudyard Kipling, and Winston Churchill. The siege of Ladysmith consisted of the British under siege by the Boers. The Boers were using the new long-range French guns of Schneider. Arthur Conan Doyle, of Sherlock Holmes fame, reported the siege:

> Huge shells—the largest that ever burst upon a battlefield—hurled from distances which were unattainable by our fifteen-pounders, enveloped our batteries in smoke and flame. One enormous Creusot gun on Pepworth Hill threw a 96-pound shell a distance of four miles, several 40-pound howitzers outweighed our field guns. And on the same day on which we were so roughly taught how large the guns were which labour and good will could haul on to the field of battle, we learned also that our enemy—to the disgrace of our Board of Ordnance be it recorded—was more in touch with modern invention than we were, and could show us not only the largest, but also the smallest, shell which had yet been used.

These great siege guns noted by Conan Doyle were Long Toms. The gun was 4.2 meters long and could hurl a 98-pound projectile more than nine kilometers. It took 30 seconds from the time the British lookout saw the white puff of smoke for the shell to hit the town. "Long Tom" inspired Krupp Works to develop a 20-mile gun, which would shell Paris in the First World War.

The military and Lincoln found agreement on the Rodman guns, but parted ways on the breech-loading cannons. The military firmly opposed Lincoln on breech-loading cannons and found many allies in the press. The actual concept of breech-loading cannons is an old one, going back to the drawings of Leonardo da Vinci. Krupp in Germany had been experimenting with breech-loaders with some success in the 1840s. Even the master salesman Alfred Krupp could not convince his Prussian generals. Lincoln remained firm in his belief in breech-loading cannons, and that ultimately brought them into use by the end of the war. The main bottleneck to successful production of breech-loading cannons was the needed American tooling and machining equipment. Some English Armstrong breech-loaders and Whitworths were purchased, which were also the only true steel cannons of the war. The Confederates, who seemed more open to breech-

Opposite: Scenes of Fort Pitt Foundry from *Harper's Weekly, 1862*

loading, purchased the Whitworths. The famous Confederate artillery general and chief of Longstreet's artillery corps E. P. Alexander was a major supporter of the Whitworth breech-loader. Other experimental designs were also used to convert Rodmans to breech-loaders at the end of the war.

Krupp in Germany was already producing breech-loaders using steel rifled barrels. Krupp had introduced steel cannons at the Great Exhibition of 1851. In the 1850s, Krupp developed the necessary tooling and machines to rifle these barrels. The development of this tooling led to the ability to produce breech-loading designs. Breech-loading mechanisms required screw threading to open and close the breech. The Union at the start of the war did have some muzzleloading steel, rifled cannons. These "steel" cannons were the 6 — and 10-pounder Wiard cannons of puddled wrought iron. Steel technology was one of the few weaknesses in America during the Civil War. Krupp had used crucible steel in his steel cannons, but of a secret composition known as "Kruppstahl." It certainly was difficult for Verne to concede this feat to the hated Krupp of Prussia. It appears in hindsight that Krupp had learned the strengthening effect of adding chromium to steel. European military generals, like Lincoln's generals, were very conservative and snubbed the use of Krupp's innovation during the 1860s. America lacked the resources for steel cannons and felt comfortable only in taking its cast-iron technology to the limit.

The Victorian hesitation concerning steel and breech-loading cannons changed with the Franco-Prussian War in 1870. Krupp biographer William Manchester described the situation:

> With the immense manpower, his hardened veterans, and his strategic frontiers, Louis Napoleon could have over come his other handicaps. France's fatal weakness was ordnance. His marshals were so preoccupied with *la glorie militaire* that they overlooked the significance of firepower. Their infantrymen's new .43-caliber, cartridge-firing chassepot rifle had twice the range of the Dreyse needle gun with some justification they expected much of their *mitrailleuse*, which, like the six-barreled American Gatling gun, was a primitive ancestor of the coming machine gun. The *mitrailleuse* had twenty-six barrels, they could be fired in rapid succession by turning a crank. But the Frenchmen's artillery was hopelessly obsolete.[3]

The French artillery was even poorer than that of America at the start of Civil War, 10 years earlier. The mainstay of the French ordnance was the bronze Napoleon. Krupp had offered breech-loading steel cannon to Louis Napoleon's representative, Leboeuf, only to be rejected in the late 1860s. In 1870, the Krupp steel breech-loaders turned the Franco-Prussian War into a rout. Louis Napoleon should have read Verne's novel! In 1873, Verne would write a political satire of the great Krupp cannon empire, *The Begum's Fortune*. Verne commented in the novel, "In gun casting, as in everything else, the man who can do what others cannot do, is sure to be well off. Indeed Schultz's cannon not only maintain an unprecedented size, but although they deteriorate slightly in use, they never burst. Stahlstadt steel seems to have special properties. There are many stories current of mysterious chemical formulae; but one thing is certain, that no one has discovered the invaluable secret." Krupp biographer William Manchester noted Verne's prediction also: "Louis Napoleon would have been better served had he consulted his most imaginative novelist. In March 1868, when Leboeuf was scrawling 'Nothing Doing' on Essen's proposal, Jules Verne was writing chapter 12 of *20,000 Leagues under the Sea*. Verne's Captain Nemo, taking Professor Aronnax on a tour of the Nautilus, explains that the submarine's engine is constructed of the strongest steel in the world, cast 'by Krupp in Prussia.'"

The Civil War inspired Victorian development and trials, which were popular events in Washington. Trials attended by officers, generals, foreign agents, and Lincoln became known as the "champagne experiments." Lincoln loved these tests and planned many himself. Historian Robert Bruce noted: "Abraham Lincoln exposed himself to that risk in his 'champagne experiments,' and more narrowly escaped accidental death. Furthermore, his risk was double, for death or injury might have come to him not only by accident, but also by design." Lincoln loved the testing and none could convince him to stay away. Lincoln read voraciously about ordnance and encouraged inventors to visit the White House. Lincoln's secretary recalled, "Specimens of new rifles and cannon came to him by the dozens with a large variety of pistols, torpedoes, and gunboats. Newly-invented guns, and specimens of all manner of old-time weapons offered for sale to the government, stood leaning against the wall in the corners, or lumbered the desks and tables." Lincoln loved the science and the testing, and initiated many of the ordnance tests himself. His own war department was more interested in standardization of arms than in applying new technology. For the Metallurgic Age, Lincoln was the American Prince Albert. Lincoln believed that technology would determine the ultimate outcome of the war. Besides inventors, Lincoln, like Prince Albert, loved the company of scientists, especially practical Victorian scientists.

The great American scientist of the period was Joseph Henry. Henry represented the model Victorian scientist, taking the basic science of others and developing practical devices. Actually, Henry stood for a new type of scientist that we know today as an engineer. Like a typical Victorian, he was a renaissance man with broad interests in science. Henry's earliest work in electricity set the groundwork for the great advances of the age. His electrical experiments had proved the possibility of electric motors and long-distance communication. In 1846, Joseph Henry took the position of secretary of the Smithsonian Institution, which he held until his death in 1878. Lincoln had held a low opinion of the Smithsonian until Henry sent him a copy of *Smithsonian Contributions*. Its pages on technology won over Lincoln, and that moved him to meet Henry. Henry and Lincoln developed a friendship that would last through the war. Henry became the "scientist royale" to Lincoln on science and technology. He even helped Lincoln expose a fake spiritualist who had preyed on Mrs. Lincoln's grief over their dead son Willie.

Lincoln later changed his opinion of the Smithsonian as well: "Professor Henry convinced me of my error. It must be a grand school if it produces such thinkers as he. He is one of the pleasantest men I have ever met; so unassuming, simple, and sincere. I wish I had a few thousand more such men." Henry was often a dinner guest with Lincoln so the two could discuss technology. The two even developed some new military technology at these dinners. One joint experiment of theirs was night signaling using calcium lights and Morse code. Henry's earlier scientific experiments had led directly to Morse's invention of the telegraph. The messages were sent between the White House and the Smithsonian tower. Another area was the development of aerial reconnaissance using balloons. In early 1861, Joseph Henry brought scientific balloonist Thaddeus Sobieski Constantine Lowe to dinner. Lincoln was amazed at the possibilities of using balloons. Robert Bruce described the meeting: "Now, as Lincoln listened, he enthusiastically sketched the possibilities of military balloon reconnaissance, even including a plan for telegraphic air-ground communication."[4]

The meeting ultimately resulted in the formation of an Army balloon corps. The balloons were stationed on a Navy vessel in the Potomac in 1862 (making it the first aircraft carrier). Lincoln put Lowe in charge of balloons, but Lowe carried no official military rank, which doomed his efforts. Balloons did contribute to the Union successes at Gettysburg and Fair Oaks, but the corps never found a niche in the military organization. Observers of the Civil War, such as Jules Verne, found other military uses for balloons. Many of Verne's novels used balloons in military applications, but in *Mysterious Island* (1874), Verne used the Civil War as background: "The government of Richmond had long since lost all possible means of communications with General Lee, and it was vitally important that the city's plight be known, so as to speed the arrival of reinforcements. It was then that this Jonathan Forester had the idea of using a balloon to pass beyond the lines of the besiegers and make his way to the separatists' headquarters." Verne actually got the idea of military balloons from the Prussian siege of Paris in 1870. The surrounded Parisians used balloons to communicate with the rest of France. The heights of the balloons prevented the Prussian riflemen from shooting them down. Alfred Krupp was called in to build a rifle that could fire directly into the air with enough firepower to penetrate the balloons. Krupp's successes with the gun made him the inventor of the first anti-aircraft gun. The Germans in general took balloon technology to its highest level. This came from a young German military observer at the Civil War battle of Fair Oaks. That observer was Count Zeppelin, who returned to Germany and started the aircraft industry.

Balloons, while novel, were not Lincoln's passion. Actually his oldest technical interest was ironclad boats. As early as 1812, inventors such as John Stevens had proposed to Congress to build ironclad ships. As a young congressman, Lincoln supported an early inventor of ironclads. Lincoln had backed an experimental ironclad of Uriah Brown, which was to be a flame-throwing ironclad. The Confederates took the lead at the start of the war with the building of an ironclad ram called the *Manassas*. The *Manassas* also had a large rifled cannon, which was a fear to the union troops up the Mississippi. At the beginning of the war, Lincoln got James Eads (future father of the steel arch bridge) a contract for the building of seven ironclads on the Mississippi. Eads already had a salvage and construction company in St. Louis. In 1861 the U.S. government contracted the building of seven "city-class" ships—the *St. Louis, Cairo, Mound City, Cincinnati, Carondelet, Louisville,* and the *Pittsburgh.* The ships were designed by Samuel Pook and became known as "Pook's Turtles." Within five weeks, Eads put the *St. Louis* into action on October 12, 1861. Eads proudly sent Lincoln a photograph and wrote, "The *St. Louis* was the first ironclad built in America. She was the first armored vessel against which the fire of a hostile battery was directed on this continent, and so far as I can ascertain, she was the first ironclad that ever engaged a naval force in the world." In general these "city-class" boats were about 175 feet long and had wooden hulls sheathed in two-and-a-half-inch-thick wrought iron.

The Confederates had no Lincoln or bountiful iron mills, but they did have a brilliant naval secretary in Stephen Mallory. Mallory had sent agents to England, France, and to the Northern states to look for potential ironclad builders. The French had the lead in ironclad technology, having experimented with them during the Crimean War (1854–1856). The French navy had launched the first ironclad, the *Gloire,* in 1859. A year later England launched the first all-iron warship, the *Warrior,* with four and a half inches of armor. Though these secret efforts failed, he did succeed in starting ironclad production at Memphis, New Orleans, and Norfolk. He ordered the destruction of railroads to supply the needed wrought iron for these projects. Mallory had pioneered an all-iron submarine, the *Hunley,* which was made from a converted wrought-iron boiler. At Norfolk the Confederates began work on a novel ironclad ship named the *Merrimac,* which was being built from the sunken hull of the Federal frigate of that name. It was to be a unique ship in that it would be completely covered with iron. It was also to be capable of sinking enemy ships by ramming. Modern ramming ships never did develop, but Jules Verne used the concept as the main weapon for his fictional submarine the *Nautilus*. In action, the *Merrimac* would sink two of the North's biggest warships via ramming. Her armor proved just as effective as cannonballs against enemy ships. The *Merrimac* would ultimately meet the world's first all-iron ship, the *Monitor*, and fight to a draw.

The construction of the *Merrimac* (known as the *Virginia* in the South) was being followed by spies and press leaks. The *Merrimac* posed a major mental challenge to the Union. The Union appointed an Ironclad Board to search out a defense. The winning proposal was from engineer John Ericsson and shipbuilder Cornelius Bushnell. Ericsson was an accomplished shipbuilder, having built the first commercial screw propeller ship, the *Francis B. Ogden*. He was a brilliant engineer but had to overcome a bias based on the USS *Princeton*. The *Princeton* was built by Ericsson in 1844 and was the most advanced warship of its time. Its novel wrought-iron cannon had, however, exploded on the gala test voyage, killing the secretaries of state and navy. The addition of Bushnell as a partner helped relieve some of the bias against Ericsson. Ericsson's iron ship design was typical of great Victorian engineering in that it took the best of the proven technology and stretched it even further.

Ericsson's design represented the first truly iron ship. The hull was wrought iron, not merely ironclad. Its design report had 40 new inventions incorporated in it.[5] These advantages were directly attributable to the superiority of the North's machine shops and tooling. It would float low in the water, only 18 inches above the water line. This low draft gave the *Monitor* the advantage of needing only 11 feet of water to maneuver, while the *Merrimac* required twice that. At the center, a two-gun rotating turret offered any angle of attack. These guns were powerful Dahlgrens. The *Merrimac* narrowed this advantage by placing a gun at each bow and four guns on each side. The *Monitor* was assembled at Continental Iron Works in Brooklyn, New York. A new type of high-pressure steam engine powered the *Monitor*. Its low profile made it impossible for solid artillery and cannon hits, with the exception of the center turret. Actually, designs for both the *Monitor* and the *Merrimac* were fluid and changing as the newspapers in both the North and South followed construction and ballistics tests. Besides the press, the Federal champagne trials and the Southern naval trials had many spies. Originally, the *Merrimac* had three layers of one-inch-thick iron armor, but as news spread of the *Monitor's* huge 11-inch Dahlgren guns, the design was reconsidered. Southern ballistics tests suggested a need for two layers of two-inch iron plate, which the South's Tredegar Foundry had many problems producing. The *Monitor* countered with eight layers of one-inch iron on the turret and five layers of one-inch plate on the upper hull. Many of the federal ballistics trials (champagne trials), were setups for Southern spies to throw off designers.

The actual battle of the ironclads was disappointing to the press because there was no clear winner. The *Merrimac* took the most, hits but most came from other amassed Union warships. The *Merrimac's* armor did have cracks and open seams, but the *Monitor's* own defects prevented it from closing in for the kill. The *Monitor's* turret overheated, with temperatures reaching 140 degrees Fahrenheit, and smoke overtaxed the ventilation system, interrupting the firing. Both ships steamed away after several hours, with the *Merrimac* requiring major rework. They never met again but played key defensive roles. The Union went on to build 20 more "*Monitors*," some with huge 15-inch guns. These ships participated in blockades but saw little combat. All of them remained in service into the 20th century, with the last one coming out of service in 1926.

The use of wrought-iron armor started a Victorian scientific competition between better armor and more powerful guns. Jules Verne had portrayed this technological battle as a basis for his novel *From the Earth to the Moon.* This competitive scientific struggle had initially inspired the work of Henry Bessemer in the development of steel making. Bessemer, in particular, was inspired by the French emperor's challenge to develop a steel cannon in response to the unveiling of Krupp's steel cannon at the Great Exhibition of 1851. The rapid development of cannons in the 1850s had motivated the British navy to study the use of iron armor. In 1857, they completed the famous "Woolwich" experiments. The British experimented with four-inch-thick wrought iron and steel armor. They discovered that four thick plates could protect against 68-pound projectiles fired at 600 yards. This was the reason to arm their first iron ship, *Warrior,* with four and half inches of wrought iron. One important finding of this study was that wrought iron made better armor than steel. The reason is that softer materials have higher impact energy. Wrought iron can therefore absorb more energy than steel. Based on information being received weekly in Britain, Sir William Armstrong started a series of experiments with armor and cannons. Armstrong proved again the efficiency of wrought iron as armor. The ironclads of the Civil War proved the point, with cannonballs bouncing off wrought-iron plate. Toward the end of the 1880s, artillerymen developed pointed, hard armor-piercing projectiles. Still, the principles of the British 1860s experiments apply even to this day. Armor today usually has a sandwich or component nature, the surface being extremely hard steel to prevent penetration of the projectile, and the backup layer being very soft steel (similar to wrought iron) to absorb the impact energy of the hit. As projectiles improved, so did armor, in an endless competition. The French and

Germans seem to have been adding chromium, manganese and maybe even vanadium to steel to improve the armor in the 1870s, giving armor the edge until the turn of the century.

Early in the Civil war, Lincoln not only promoted armor/cannon trials, but also nurtured an interest in the use of body armor. The Confederates had tried some use of body armor but without success. One of the failed Confederate armor breastplates was displayed in a prominent Washington jewelry store. There were still many individual claims of success. The 14th New Hampshire Regiment tried a Union experiment in October of 1862. The failure of this experiment was recorded by the regiment historian: "The first halt was made in front of the White House, and at least one-third of the battalion took a vigorous account of stock. The men with bullet-proof vests—their hope and pride—in Concord ... vowed they would prefer to risk rebel bullets rather than carry so much old iron any further. Steel breastplates sufficient to coat a small gunboat were hurled into the gutter in front of Father Abraham's marble cottage."

The Civil War defined the Metallurgic Age, bringing advances in ferrous metallurgy, casting methods, and machining. Before the war, Cleveland had a handful of foundries After the war, it had more than 80! Pittsburgh, a minor producer of steel prior to the war, rose to dominance in the world after the war. Machine shops were few prior to the war, while afterwards we were a nation of machinists. Where Britain and France had dominated the industrial shows prior to the war, America dominated after. Industrial management and quality control had their birth during the Civil War. The telegraph went from a few hundred miles of wire to tens of thousands after the war. The news of Fort Sumter (the start of the war) took more than four weeks to reach London, while the news of Lincoln's death took twelve days. At the start of the war, bronze cannon hurled 20 — to 30-pound shells a few hundred yards; after the war a 100-pound shell could be shot from miles away with monstrous iron cannons. Prior to the war, a well-trained soldier could fire three rounds a minute from his rifle, while after the war he could fire as many as 50 in a minute. Prior to the war it could take weeks to move an army 100 miles after it could be done in a day. A general would need hours to deliver a message to the front prior to the war. After it could be done in seconds. Wood was used for ship armor prior to the war. Afterwards, it was wrought iron. The number of charcoal furnaces in the United States went from fewer than one hundred to more than a thousand. The world moved from one of cellulose to one of iron, transforming bridges, ships, and railroads.

7

Trains and Boats

It was not war that drove the Victorian economic engine; it was the steam locomotive. The growth of the railroads and general industrial growth became interrelated. The steam locomotive was not a Victorian invention; its roots go deep into the 18th century. However, the history of the steam locomotive is beyond the scope of this book. The year 1800 stands out as critical, however, because in that year James Watt's patents on the steam engine ran out, freeing inventors. We will start our historical inquiry here, at the beginning of the 19th century. By the dawn of the century, there had been a number of steam locomotives in operation, some running on rails but most capable of running without rails. When rails were used, they were of wood, with some cast iron. In 1801, the first recorded transportation of passengers by the force of steam was recorded.[1] This first "train" did not use rails, but rather wood plates or grooved stone were used to distribute the weight of the engine. Richard Trevithick of England designed this innovative locomotive. Trevithick's engine operated at low pressure, common in the 18th century because of the lack in quality iron and steel plate. Low pressure limited the hauling power and speed of the locomotive. The operation was merely a historical experiment, never achieving commercial reward. Still, Trevithick succeeded at applying a higher-pressure locomotive within a year and added rails of wood and stone in 1803. Trevithick then designed the first steam locomotive on rails. His locomotive carried 10 tons of iron a distance of nine miles in four hours and ensured the future codependency of the railroad and iron industry.

The development of the steam railroads continued with a number of successes and failures, but nothing of any commercial value emerged until 1828. In 1828, the Baltimore and Ohio Railroad was formed. The man who opened up the railroad business for the Baltimore and Ohio Railroad was the inventor of the first lawnmower and iron master — Peter Cooper. Cooper put his first locomotive on tracks in 1830 to carry iron ore but soon turned to passengers. The famous "Tom Thumb" of Cooper ran 13 miles in 70 minutes carrying 42 people. These first rails of Cooper's design were of stone and wood. Cooper, being a Victorian engineer, knew the importance of the press and technology in any commercial success. Cooper promoted a "horse race" to show that the railroads had come of age. Cooper's locomotive failed to live up to the promotion; "Tom Thumb" lost the race to a horse pulling the same load on a parallel track. Still, the publicity helped acclimate the public to the idea of railroad travel. The Baltimore and Ohio did in 1830 become the first railroad to open for paying passengers, routinely traveling 13 miles around Baltimore. By 1835, more than 1,000 miles of track had been laid. Cooper's advantage came from the superiority of his wrought iron, which was the best boilerplate available. Cooper was a brilliant industrialist and railroad investor who had built a very successful ironworks in Trenton, New Jersey. To supply iron ore from the nearby mountains, he built his own railroad. 1830 stands out as a key year for railways, with a pioneer railroad in Austria opening with 80 miles of cast-iron track. Also that year, a 31-mile Manchester and Liverpool commercial railway was opened in England; these rails were of wrought iron. The Manchester and Liverpool Railroad would debut its famous locomotive the "Rocket," built by Robert Stephenson. The Manchester and Liverpool Railroad had a publicity contest of its own, offering a $2,500 prize for the fastest locomotive. The "Rocket," in winning the competition for the Manchester and Liverpool Railroad, set a new speed record of 30 miles per hour. One of the losing entries was by the future builder of the Union's iron boat, the *Monitor*, John Ericsson. This year of 1830 would start the competition among wood, cast iron, and wrought iron for rails and locomotive wheels as well.

That material competition for rails would last until the 1880s, when steel won the battle. If you are not an engineer, you may first ask why rails at all. The function of the rail is to distribute the weight. Without rails, the heavily loaded locomotive and car wheels would sink deep into the ground. This would become even more pronounced on soft ground or wet ground. Wood and stone "plates" were at first popular because of the ease of manufacture and the availability of wood. Wood, however, had a short life and low load-bearing properties. In 1832, there were 67 railroads in Pennsylvania varying from a few hundred miles to 23 miles, and they were predominately wood. Stone seemed a popular alternative, but cast iron and wrought iron offered the strength needed. Inventors such as Peter Cooper became supporters of wrought iron because of their understanding of material properties. Cooper, of course, had a huge interest in iron as a manufacturer. Wrought iron was superior to cast iron in that it was flexible and could handle the impact of a fast-moving trains while cast iron would fracture under impact loading. Still, if the rails were linked smoothly, cast iron would handle large compressive loads. For this reason it was favored by railroad builder Robert Stephenson. Wrought iron did have one drawback: On hot days with heavy loads, the rails would "snake." Snaking caused the wrought-iron rail to curl up and cause wrecks. Rails required the additional strength of steel, but steel was in very short supply until 1870. This problem led Cooper to produce the first Bessemer steel at his Trenton plant in 1856. During the Civil War, steel rails started to appear. In 1870, only 5 percent of the rails were steel, with the balance predominantly wrought iron.[2] By 1890, the mix was reversed in favor of steel. That 20-year period was one of fast technological advances, which frustrated and bewildered investors.

Bessemer steel started its climb to dominance in the rails in 1875 with the opening of Andrew Carnegie's first steel plant in Braddock, Pennsylvania. Carnegie had gained much experience as a manager of the Pennsylvania Railroad in the 1860s. This included a great deal of technical expertise. The president of the Pennsylvania Railroad, Edgar Thomson, was an early convert to steel rails. At the time, Carnegie was working with him. The Pennsylvania Railroad had test sections of steel trials that were showing exceptional wear results. Thomson asked Carnegie to make a trip to Europe in 1865 to study the use of steel rails as well as the Bessemer steel-making process. Carnegie even discussed the use of steel rails with Henry Bessemer. A British patent of Thomas Dodds for capped welded steel to wrought-iron rails also captured Carnegie's interest. The product offered a tough, flexible wrought-iron base with a hard, wear-resistant surface. It was the same technology used centuries before by Samurai sword makers, who forged swords with soft, flexible iron centers but hard steel surfaces for cutting. Carnegie returned to America to push the Dodds process rails. Some field failures did slow Carnegie's enthusiasm and ultimately formed his business strategy — "Pioneering doesn't pay." The Dodds process failed because of the lack of welding technology. It would, of course, be one of Carnegie's rare failures. Later, when a manager, Tom Scott, at the Pennsylvania Railroad asked Carnegie about investing in "chrome" steel, Carnegie replied, "My advice (which don't cost anything if of no value) would be to have nothing to do with this or any other great change in the manufacture of steel and iron. I know at least six inventors who have a secret, all are anxiously awaiting.... That there is to be a great change in the manufacture of iron and steel some of these years is probable, but exactly what form it is to take no one knows. I advise you to steer clear of the whole thing. One will win, but many will lose & you & I not being practical men would wager long odds. There are many enterprises we can go in even."[3] Carnegie hesitated even as Edgar Thomson's enthusiasm for all-steel rails grew. Carnegie's brother and advisor, Tom, was convinced early on of the future of steel rails. Ultimately, the future of steel was set by the rapid spread of the Bessemer Process throughout Europe between 1865 and 1872.

Carnegie continued to watch and study the technology of steel making and application. In 1873, based on his own studies, the trials of the Pennsylvania Railroad, his partners' convictions, the suggestion of steel engineer Alexander Holley, and the growing market signs, Carnegie moved into steel rails. He moved to finance a steel mill dedicated to the production of Bessemer steel rails. With the decision made, he successfully opened his plant at Braddock in 1875, naming it after Thomson. Carnegie's timing was perfect, as the first Bessemer plants and metallurgists such as Robert Mushet eliminated process problems and fine-tuned the process. Within 10 years, Carnegie owned the steel-rail market and wrought iron was pushed out. The railroads and the Carnegie steel empire became codependent, with Carnegie's biggest customer being the railroad and the railroad's biggest customer being Carnegie. Steel production required almost continuous railroad cars of coal and iron ore.

As Bessemer steel became more available, it made inroads into all locomotive parts. The first use of steel in locomotive fire boxes in England in the 1850s failed because the heating caused the steel to harden and crack. The Pennsylvania Railroad finally was successful in building steel fireboxes in 1862. Wheels were also an area of ferrous metallurgy evolution. In the 18th century cast-iron wheels and wood were in vogue. The first forged wrought iron wheel came into being in 1835 in England. Wrought iron tended to be too soft. Railroad designer Robert Stephenson preferred cast iron and patented the cast-iron wheel with spokes. Cast iron was a great material because of its ease of manufacture and its natural compressive strength (ability to handle heavy loads). One problem, however, was that cast iron as casted is a "gray" iron. Gray iron has large graphite flakes in the microscopic structure. Gray iron machines well but wears poorly because of these graphite flakes. The reason is that these graphite flakes act as chip-making points in machining and lubricate the tooling in application. Since machining exemplifies accelerated wear, gray iron wears in rail service by the same graphite mechanism. The use of the term "gray" refers to the color of the iron when fractured. English foundrymen discovered in the 1830s that chilling the outer rim of the wheel produced a harder, more wear-resistant wheel surface. This chilled or "white" iron has hard carbides instead of graphite. The carbides are hard and very wear-resistant. The chilled wheel combined a strong, load-bearing gray-iron center (hub) with a hard, wear resistant-rim (tire). The chilled-iron concept is still important today in many manufacturing and wear applications, such as brake linings. The problem with chilled carbide railroad tires arises from the need to be matched to rails in hardness. This was not understood in the 1800s, and chilled carbide wheels caused excessive wear of the softer gray iron or wrought-iron rails. Wear minimization theory requires the wear surfaces to be matched, or else the softer surface will wear rapidly. Wheels and rails should be of the same material and processed to similar hardness levels. It was the same problem that modern engineers discovered in the 1970s, when carbide-studded rubber tires for winter driving were chewing up the nation's softer cement highways.

Steel wheels were developed in 1870 by "tiring" or "steeling" cast iron wheels following similar developments in rail manufacture a few years prior. They became known as Washburn wheels, made by pouring cast iron into a flask with a steel rim. This steeling of cast iron evolved from a blacksmithing technique used for centuries where steel was in short sup-

Tensile testing of steel for locomotive applications from *Scientific American 1890*

ply. By 1880, railroad engineers started to see the superiority of steel rails and wheels. Forged steel wheels became popular because of their toughness and improved wear. As early as 1849, Krupp Steel had supplied two steel axles to the Pennsylvania Railroad. The test showed almost no wear in five years after 80,000 miles of use. Krupp's success with steel axles gained it many market, inroads including the use of steel in locomotive cylinders. Krupp also became a major supplier to the Atchison, Topeka and Santa Fe Railroad in the West and the New York Central in the East. In the 1890s steel axles started to replace wrought-iron axles, as steel became king. Steel would rule from 1890 on.

By the early 1820s, the main deterrent to the advance of railroads was the need for bridges to carry the weight of the locomotive and train. Robert Stephenson would be the man to overcome that obstacle. Robert was the son of George Stephenson, who had built the first commercial railroad from Liverpool to Manchester. George and Robert had revolutionized bridge materials and design. Robert's "tubular" bridges opened up the continents to railroads. He not only pioneered new designs, but also applied new technology such as steam-driven hammers to drive piers for the bridges. As we have seen, wood was replaced by cast iron, cast iron replaced by wrought iron, and finally wrought iron by steel. Along with the improvement in metallurgical materials came ever-increasing loads that could be transported. The transportation of people was the most important factor in the development of the railroads, but the comfort of people in this new technology did not improve until the 1850s.

Charles Dickens, a critic of industrialization, recorded these thoughts about American rail service in an 1842 visit to the United States: "There are no first and second class carriages as with us; but there is a gentlemen's car and a ladies' car: the main distinction between which is that in the first, everybody smokes; and in the second, nobody does. As a black man never travels with a white one, there is also a Negro car; which is a great blundering clumsy chest.... There is a great deal of jolting, a great deal of noise, a great deal of wall, not much window, a locomotive engine, a shriek, and a bell.... The cars are like shabby omnibuses.... In the center of the carriage there is usually a stove ... insufferably close." This was an improvement over the 1830s, when there was a fear of traveling on trains. In addition, local communities resisted the laying of track through their communities. There was a "right side" and "wrong side" of the tracks. The wrong side was where the side of the prevailing winds blew the soot, dust, and smoke into the neighborhood. The smoke, dirt, and smog were the price of metallurgical prosperity. Cities like Pittsburgh, where train smog combined with coal-burning and steel-production smoke, experienced artificial nights, with streetlights being required by early afternoon. Writers such as Dickens wailed that the price of this prosperity was too high. Environmentalists would have been outraged by the impact of this air pollution, which even changed local weather. On cold winter nights the smoke and huge moisture releases from steel mills in the Pittsburgh area would actually create snowfalls in nearby neighborhoods! Still, the growth of the railroads was an inherent part of the Metallurgic Age.

America was being drawn westward by free land and the finding of gold in California and silver in Nevada. The paths to the West Coast of America were all problematic. One option was to take the 16,000-mile, two-month sea journey around Cape Horn in South America. Another sea option, while faster (six weeks), was extremely expensive. This option included a sea journey to Central America followed by a land trek through the jungles, rife with malaria and cholera, only to board another ship for the final leg of the trip. With luck, you might make California in five weeks. The preferred option for the average traveler was by wagon train, which took three to six months and included many dangers. One inventor, Rufus Porter, even sold seats in his planned steam-powered dirigible to make the trip in a day. Porter's dirigible was never built even though many signed up for the trip.

Besides gold, two political movements drew people as well to California. First was the popular notion of "Manifest Destiny." Land was cheap or often free to the first settlers. Second, the Northern states desired to add antislavery states to the Union. The idea of a 3,000-mile railroad made political sense, even without considering the commercial factors, such as linking the factories of the East with California. A California railroad builder in the 1850s understood the importance of linking America's industrial might. In 1854, during the Gold Rush, Theodore Judah built the 21-mile Sacramento Valley Railroad. The cost was enormous, maybe as high as $100,000 per mile of track. Part of the reason was that he had to ship wrought-iron rails and heavy locomotives around Cape Horn by clipper ships. Even much-needed dynamite had to make the trip around Cape Horn. Several of these dynamite shipments resulted in large ship explosions. Midwest merchants also dreamed of bringing Asian goods to market via a West Coast rail connection.

By the early 1850s, the idea of a transcontinental railroad evolved into a national dream. In 1860, Abraham Lincoln had made a campaign promise to start a transcontinental railroad. Theodore Judah became passionate about the commercial advantages of a transcontinental railroad. In 1861, Judah put together some prosperous West Coast merchants to form the Central Pacific Railroad. Judah became chief engineer and spokesman. Even with the Civil War, Judah had no problem convincing Congress of the project's need. The Pacific Railroad Act of 1862 passed, forming a new company — the Union Pacific Railroad Company. The Union Pacific would build west from Omaha, and the Central Pacific would build east from Sacramento. One interesting side note was that the law required the use of American iron even through British iron was available on the market at a third of the price. Nationalism and industrial pride were the basic attributes of any Victorian dream. This was the type of Victorian dream that inspired the best of the capitalists, engineers, and politicians. Most important, it would advance technology and offer a challenge. The Sierras alone would require 15 tunnels and would ultimately take five years to cross.

Tunneling of any length had to wait because of the lack of tools such as drills and dynamite. The series of transcontinental tunnels had started in the era of massive manpower and hand picking. In Europe, an eight-mile tunnel had begun at Mount Cenis in 1857. An Italian engineer, Germain Sommeiler (1815–1871), improved on some earlier steam drills by using compressed air. His drill is said to have "proved to be the key that unlocked the stone portals of Mount Cenis." Still, Mount Cenis required 15 years and several other inventions such as dynamite to be completed. The technology used on the Mount Cenis was transferred quickly to use on the Transcontinental Railroad. This new technology included the use of compressed-air power drills. History hails the appearance of compressed air, but the real advance was the steel drill. High-quality tool steel became available from American, German, and English steel makers. High-speed steel arrived by the steady improvements of crucible steel makers, not the headline advances of Bessemer steel makers. Drill steel is high-carbon steel and has high strength compared with the lower-carbon Bessemer steel.

The 1,659-foot Summit Tunnel of the Transcontinental Railroad is still today considered the most expensive railroad tunnel ever built. Tunneling applied the use of nitroglycerin, newly invented steam drills, and massive numbers of tenacious Chinese laborers. Tunneling progressed at 7 to 12 inches per day.[4] Legendary John Henry, a black steel driller from West Virginia, did successfully challenge the new steam drills with a 30-pound hammer, but there weren't enough Henrys to improve the pace. Nitroglycerin presented its own problems. Invented in 1846 by Italian chemist Ascanio Sobrero, it had three times the power of gunpowder. The problem was that it was extremely unstable. The Union Pacific turned to it because of its power, but there was almost no practical experience in its dangerous use. Many men were lost in learning to handle it. Transportation of nitroglycerin challenged the best of the civil engineers. One shipload of it exploded en route from South America, killing more than 50 men. Special tin and rubber rail cars were invented to carry it. The great loss of life inspired another Victorian, Alfred Nobel, to invent a more stable form of "explosive." Dynamite came too late for the Sierra tunnels. Dynamite would, however, not only finish the tunneling job at Mount Cenis, but also initiate the nine-and-a-half-mile-long St. Gotthard tunnel in Europe. The St. Gotthard also Tunnel had the advantage of pneumatic drills. The St. Gotthard Tunnel progressed at a mile a year versus a half mile for Mount Cenis.[5] Dynamite did not overcome the problems of poor ventilation, unsanitary conditions and high temperatures, which resulted in more than 600 deaths at St. Gotthard.

Like the use and handling of nitroglycerin, the main technological advances of the Transcontinental Railroad were hundreds of small inventions and civil engineering techniques. Even as the railroad progressed beyond the mountains, it faced new technological challenges. One very basic one was the lack of hardwood on the Great Plains. Hardwood was critical for good railroad ties. The plains had mainly softwoods, such as the abundant cottonwood. A temporary solution was to use the invention of a British chemist, Sir William Burnett. Burnett had developed a process that chemically treated cottonwood ties with zinc chloride. Another improvement from the Transcontinental Railroad was incentive pay. A labor boss, Jack Casement, found a way to motivate the laborers with a pound of tobacco to each man for every extra mile of track laid in a day, and two dollars to each man for four extra miles. On Central Pacific's best day on the plains, 1,400 men laid 10 miles of track. The two companies met on May 10, 1869, after six years of work, but seven years ahead of the original schedule. Telegraphs sent the message around the country. The following describes the celebration: "While photographs of this historic event were being taken in Promontory, the rest of the nation was celebrating. In Washington D.C., a magnetic ball on top of the Capitol was dropped to signal the beginning of the public celebration. In Buffalo, thousands of people poured into the streets to sing The Star-Spangled Banner. In New York City, soldiers fired a 100-gun salute, and Wall Street closed for the day. In Chicago, people jammed the streets in a parade that was seven miles long. In Sacramento, 30 locomotives blew their whistles and rang their bells as thousands cheered. In San Francisco, cannons boomed far into the night."[6]

Within five days of the meeting, the railroad was fully operational. The trip from Omaha to Sacramento now took less than a week. In the first year the Transcontinental Railroad moved 150,000 passengers. The Pullman Palace Car Company that supplied luxury overnight travel addressed Dickens's complaints of rail travel in the 1840s. Pullman cars were designed with plush interiors, servants, expensive furniture and fresh daily linens. The lowest-class travelers, however, had only wooden benches at a cost of $40. For $80 a traveler could have upholstered chairs, while the Pullman luxury car cost $100. This price did not include food, which was supplied by eating-houses at stops along the road. Actually, these half-hour dining stops did slow the trip. Dining cars did not appear until 1875.

Central to the spread of railroads was the development of bridges. We have seen the material technology of bridges advance from wood to stone to cast iron to wrought iron and finally to steel, but along with the materials, the designs advanced. Historians of bridge building, such as Henry Petroski, place the start of the modern bridge-building era at the end of the 18th century. The modern era begins with French engineer Jean Rodolphe Perronet (1708–1794), who advanced the basic technology of the Roman stone arched bridge to multi-arch spans. Perronet applied a basic engineering principle unknown to earlier bridge builders. That principle is based on the fact that in multi-arch bridges, the intermediate arch piers do not carry the thrust of the full bridge but support only the dead weight of the individual arches. The abutments on either bank carried the thrust of the bridge. The intermediate piers would be slender so as not to be major obstructions to water flow. In 1787 King Louis XVI of France requested Perronet to build a stone, multi-arch bridge over the Seine in Paris. It was built from stone generated by the French revolutionaries tearing down the Bastille. This famous bridge is known in France as Pont de la Concorde. It was originally to have borne the name of Louis XVI, but he went to the guillotine the year of its completion. The famous Abraham Darby's (1750–1791) cast-iron bridge at Coalbrookdale, England, built in 1779, was also a basic arch.

Scottish bridge builder Thomas Telford integrated the properties of cast iron with the engineering characteristics of the arch, inventing the truly modern iron arch bridge. In 1795 Telford was asked to replace an old stone bridge. Originally he started to build using the same approach of stone blocks except replacing them with solid cast-iron sections. He experimented as he built, discovering that the arch could be made of cast-iron ribs and still carry the full load of the bridge. Telford also realized that the cast iron in arch bridges was in compression. Cast iron excels in compressive strength beyond stone, wrought iron, and even steel! All the strength of cast iron is in compression; in tension (pulling-type stress) cast iron is brittle and weak. Few engineers understood cast iron's unusual strength characteristics, but Telford used this metallurgical principle throughout his career. His magnum opus was to be the Menai Straits Bridge in Wales (1825). This bridge was to be a suspension bridge but was a perfect application of two materials—cast iron and wrought iron. Telford applied wrought iron in the chains used to suspend the bridge because of the tensile strength of wrought iron. Suspension bridges have flexibility, but these early suspension bridges were not well-adapted for the heavy loads of railroads. Telford's legacy would be the harmony of design between material characteristics and engineering principles.

It would take an American railroad engineer to further refine the bridge design principles for the railroad industry. Squire Whipple (1804–1888) deserves the title of "father of American bridges" because of these refinements. Whipple patented an arch-truss bridge in which the cast-iron top parts and chords were always under compression and the wrought iron lower chords were always in tension. Whipple advanced the science of bridge building with the use of mathematical analysis of the stresses and strains. Whipple was then able to be economical in his quotes as well as predict actual safety factors, which would be applied by the next generation of builders. Whipple was a tireless promoter of wrought iron in railroad bridges because of its tensile strength and fire resistance. These sturdy iron bridges were ideal for short railroad spans. In addition, Whipple went on to publish many engineering articles on bridges as well as a classic definitive book, *Elementary and Practical Treatise on Bridge Building*.

The arch bridge remained popular with the advent of steel, which added inherent material strength to the design. German engineer Charles Pfeiffer had used a modified "steel" arch in the building of the Koblenz Bridge. Pfeiffer had won several awards in Germany for his theory of arch bridges. One of Pfeiffer's associates, Henry Flad, helped with the Koblenz Bridge before immigrating to the United States. Pfeiffer and Flad both advanced the use of drafting and the use of models in bridge design. During the Civil War, Flad gained valuable railroad building experience. In 1869, Flad teamed up with James Eads, and they were issued a patent together for "an improvement in arched bridges." Pfeiffer and Flad were reunited under the supervision of Eads in building the famous St. Louis steel-arch bridge, the Eads Bridge, in the 1870s. This bridge opened up railroad transportation to the West. The bridge was the first to use high-strength chromium steel. It included a unique arch made up of cast-steel tubular segments. Maybe one of the greatest contributions of the Eads Bridge arose from the manufacturing and construction processes. Eads implemented a strict process control and inspection from steel manufacture to construction practices. The Eads Bridge promoted steel usage in bridges worldwide. Eads had to overcome the argument of Andrew Carnegie's bridge engineers that a steel arch could not span the needed 600 feet. Eads believed in the famous calculations of Thomas Telford, who said a cast-iron arch could do 500 feet. Eads figured the additional strength of steel would allow for 600. Today steel arches span 1,600 feet. The St. Louis Bridge stands as a great engineering success. Eads went on to become the first American to win the Albert Medal given by the Society for the Encouragement of Art, Manufactures, and Commerce. Eads represents one of the Victorian meta-inventors who applied his creativity in the diverse areas of engineering, such as bridge building, shipbuilding, construction, and quality engineering.

The design of railroad bridges advanced with the terrible Tay Bridge disaster of 1879. On December 28, 1879, the Tay Bridge in Scotland (also known as the Firth of Forth Bridge) failed as a train from Edinburgh passed over. All 70 passengers on the train died. The bridge at the time was the longest in the world, spanning almost two miles of water and having a bend in it. The design consisted of a truss-and-girder structure that had a successful record in short span applications. The builder and designer, Thomas Bouch, was knighted for his design a year before its failure. In hindsight, the design lacked the wind resistance needed for its location, and on the day of failure, the bridge was in the middle of a gale. Because of the number of deaths involved, a highly public commission reviewed every facet of design and construction. The cast-iron members came under particular scrutiny and were discovered to be of poor manufacture. The cast-iron columns suffered from casting defects known as blowholes. These are large voids in the cast iron that significantly reduce the strength of the girder or column. The Regulation of Railways Act of 1871 did require on-site bridge inspection, but it appears the foundry had cosmetically filled the visible holes with a batching paste. The paste, of course, added nothing to the strength of the overall member. The commission issued its findings: "Sir

Thomas Bouch is, in our opinion, mainly to blame. For the faults of design he is entirely responsible. For those of construction he is principally to blame in not having exercised that supervision over the work, which would have enabled him to detect and apply a remedy to them. And for the faults of maintenance he is also [principally], if not entirely, to blame in having neglected to maintain such an inspection over the structure, as its character imperatively demanded." Compare this to a few years earlier with the building of the Eads Bridge, the almost obsessive inspection and testing by Eads. Eads's inspection and testing techniques would have prevented the failure.

The replacement of the Tay Bridge brought the best engineering of the period together. The new designers were to be John Fowler (1817–1898) and Benjamin Baker (1840–1907). Fowler had become known somewhat as a seer, in that he forbade his family to travel across the Tay Bridge. They started their project with a detailed study of the stresses on the bridge using models. Baker's study suggested that the winds posed a greater problem than the weight. Baker proposed a "new" design and engineering principle, which would become known as the cantilever bridge. Eads had really pioneered the idea of cantilevering in his Eads Bridge in St. Louis. Cantilever design was evolving throughout the world. Six weeks prior to the Baker and Fowler proposal using the cantilever principle, a British-trained Japanese engineer, Kaichi Watanabe, proposed it in *Engineering News*. In 1883, an Eads "apprentice," Charles Schneider, at the Niagara Gorge, built a successful cantilever railroad bridge. The successes of the Schneider Bridge had proved the feasibility of using a cantilever design in different applications. The public scrutiny of the new Tay Bridge required Baker to defend his new proposals in lectures and technical papers. He reviewed the risks of even minor factors, considering the uneven effects of sun heating on the bridge expansion. Probably the biggest improvement of the bridge was metallurgical. The original Tay Bridge used cast-iron members in tension. As we have discussed, cast iron under tension or twisting stress created by wind is extremely weak. Baker proposed to use steel: not Bessemer steel, but technically superior open-hearth (Siemens-Martin process) steel. Open-hearth steel is "cleaner" and therefore much stronger than Bessemer steel. The bridge was completed in 1890 and is considered the transition point from wrought iron to steel in world bridge building. The new Tay Bridge would become the first open-hearth steel bridge and maintain its integrity to this day. Construction of this unique bridge included the use of electrical arc lamps for lighting.

Another one of the engineers affected by the Tay Bridge failure was Theodore Cooper. Cooper by 1880 had risen to the pinnacle of bridge experts. Cooper had worked with Eads on the Eads Bridge and John Roebling on the Brooklyn Bridge. He managed the Keystone Bridge Manufacturing Company in Pittsburgh. It was his early career as an inspector for James Eads that defined him. Eads sent him to Midvale Steel and the Keystone Bridge Company to monitor and inspect steel bridge members. Cooper checked that the components met the rigid standards of the design. Cooper oversaw the testing of the steel as well as set the restrictions. With Cooper's help the Eads process emerged as a model for bridge construction and national bridge standards. In 1885, Cooper published his first book, *General Specifications for Iron Railroad Bridges and Viaducts*. The book was hailed as the "first authorative specifications on bridge construction." Technically, Cooper surpassed his master, Eads, in the understanding of materials and their applications. Cooper became a pioneer in the design of steel bridges, having published a medal-winning engineering paper, "The Use of Steel for Railway Bridges."

Cooper stood alone at times with his emphasis on economic bridge construction. He was one of the first to suggest the concept of a "safety factor" in bridge design. Cooper proposed what came to be called the "American system of competitive bridge building." Engineering historian Henry Petroski highlighted Cooper's view:

> Cooper believed that bridges should be made strong enough to perform their function but not so strong that they are heavier and more expensive than they have to be. This view has always been and always will be shared by the best of engineers, for they recognize that in the final analysis engineering is part of a much larger social enterprise, and money spent unnecessarily for civil-engineering structures becomes unavailable for other civic endeavors, or for initiatives of a humanitarian kind. The line between too little and too much safety is not always very clear and distinct however, and that is what makes the best engineering also the most difficult. It is also what can lead to the worst engineering.[7]

Cooper's economic engineering principle became a Victorian paradigm that exists to this day. It is applied in shipbuilding, aircraft, automobiles, and buildings as well as bridges. Cooper's economic approach favored the cantilever design, but another design was gaining popularity.

Suspension bridges, such as the Brooklyn Bridge and the Niagara Gorge Bridge, were dominating the Victorian headlines. The suspension bridge gained fame through the works of John August Roebling (1806–1869) and his son, Washington Roebling (1837–1926). John Roebling invented a wire-twisting machine in the 1850s that produced wrought iron cable. Using the examples of primitive rope pedestrian bridges, the elder Roebling envisioned a suspension bridge having the strength of wrought-iron cable. In 1856, Roebling constructed the first suspension bridge across the Niagara Gorge. The bridge spanned 821 feet over the scenic gorge, capturing the imagination of any observer. Roebling erected suspension bridges in Pittsburgh and Cincinnati prior to his greatest achievement, the Brooklyn Bridge. The Brooklyn Bridge would be called at the time "the Eighth Wonder of the World." His use of crucible steel wire and Bessemer steel girders shocked engineers in 1878, but had been inspired by Eads. Actually, Roebling unknowingly was the first to use Bessemer wire. Roebling's wire specifications for suppliers required the use of crucible steel to meet the standards. The problem was the suppliers could not produce enough crucible steel to meet the volume and substituted Bessemer steel. Some Bessemer steel did get into the bridge wire even though Roebling had restricted it. Roebling pioneered a number of inventions with the Brooklyn Bridge. Galvanized (zinc-coated) wire was applied for the first time in suspension bridges. Roebling used a truss-type approach to the roadbed of the Brooklyn Bridge to increase its rigidity. Suspension bridge disadvantages include some aerodynamic instability and a tendency for vibration stresses. The bridges dominated the end of the Victorian Metallurgic Age, as the great ships had stirred the Victorian imagination earlier in the century.

Shipbuilding defined a nation's engineering and technical might. The great Victorian essayist Ralph Waldo Emerson described it best in his essay on American civilization:

> The ship, in its latest complete equipment, is an abridgement and compend of a nation's arts: the ship steered by compass and chart longitude reckoned by lunar observation, and, when the heavens are hid, by chronometer; driven by steam; and in wildest sea mountains, at vast distances from home, "The pulses of her iron heart, Go beating through the storm." No use can lessen the wonder of this control, by so weak a creature, of forces so prodigious. I remember I watched, in crossing the sea, the beautiful skill whereby the engine in its constant working was made to produce two hundred gallons of fresh water out of salt water, every hour, thereby supplying all the ship's want.

The use of wrought iron in ships began long before the Civil War. The first use of iron in shipbuilding occurred in 1818 with the launching of the *Volcano*. The masterpiece of the Metallurgic Age became the iron steamboat. Isambard Kingdom Brunel (1806–1859) was the Victorian that made iron king in shipbuilding. Isambard was the son of another successful engineer, Sir Marc Isambard Brunel (1769–1849). Marc Brunel had been a French royalist who escaped to New York during the French Revolution. In 1796, Marc became engineer for New York City and had many successful, projects including the building of a cannon factory. In 1799, he returned to England with a new invention. The invention was a tackle-and-pulley system for loading and unloading ships. The pulley mechanism would require some complex machining and machine tools that were not available locally. America, though short in this expertise in 1800, would later excel in it. Marc

Brunel teamed up with British toolmaker Henry Maudslay, who would become known as the father of machine tools. With Marc moving to England to work with Maudslay, America lost a great inventor, designer, and engineer. Maudslay developed a set of tools to produce the forty-plus parts that Brunel's system required. Finally, Marc's, hoisting system was installed in the Portsmouth Navy Yard in 1808. Success was immediate, and he got large royalties from the British government. In 1828, Marc started his greatest engineering feat, the Thames River Tunnel, which his son, Isambard, would complete.

Isambard Kingdom Brunel, while educated in France, joined his father's construction company to learn the trade. Isambard eclipsed his father's success in bridge building, having invented the cantilever-type design. At age, 27, he became chief engineer for the Great Western Railway Company. In that capacity he took an interest in building a transatlantic steamboat to further the logistics network of the company. He convinced the company of the project's feasibility in 1835. Brunel's ship, the *Great Western*, would be the longest steamer of its time at 212 feet. It was a side-wheel steamship with a unique steam engine built by his father's old friend Henry Maudslay. In its maiden voyage in 1838, the ship crossed the Atlantic in a record time of 15 days. Actually, the *Sirius* crossed the Atlantic a few days prior to the *Great Western*. Still, the *Sirius* had to burn its cabins and woodwork for fuel after running out of coal. The *Great Western* became the first transatlantic steamship for routine service. The ship's success propelled Brunel into fame throughout Britain and Europe.

Brunel's second ship was an iron-hulled steamboat, the first iron steamboat ever built. The *Great Britain* was 322 feet long and made its maiden voyage from Liverpool to New York in 1845. Probably the major engineering feat of the *Great Britain* was its use of a screw propeller instead of side paddle. The screw propeller provided the efficiency necessary to cross the Atlantic with ease. Side paddles ships could barely cross because of the coal needed for the voyage. Jules Verne years later described the common occurrence of paddle wheelers having to burn materials on board for fuel in *Around the World in Eighty Days*. Still, even with the efficiency of a screw propeller, the *Great Britain* could not carry enough coal for the important Britain-to-Australia route. It should be noted that no ports for refueling existed in the 1840s, requiring self-sufficiency. Self-sufficiency on the Australian route would be the motivation for Brunel's *Great Eastern* and ensured investment money. The huge *Great Eastern* would be twice the length and 10 times the displacement of the *Great Britain*. The sheer size of the ship presented a number of engineering problems. The biggest engineering feat was to launch it sideways from 100 yards inland. The rigging devised by Brunel needed daily corrections. The launch moved only a few inches per hour, taking three months to complete. The *Great Eastern* was a combination paddle wheel and screw propeller capable of the long Australia voyage. The *Great Western* and *Great Eastern* have been lost to the scrap dealer, but the *Great Britain* is restored in dry dock at Bristol Harbor. The *Great Britain* was never a commercial success because of mechanical problems with its novel screw, and navigation-compass problems attributable to the iron hull. Another advance of the *Great Eastern* was the use of engineering modeling in the design. Brunel consulted an old railroad associate of his, William Froude (1810–1879), to help in the design. Froude experimented with wax models to check various designs. These models were three to six feet long. He used a large test tank, which is common today to test ship designs. The British navy was favorably impressed with Froude's test tank and contracted him to build a similar one for the Admiralty. Froude used testing equipment to measure the resistance of hull designs as well. Ultimately, Froude developed formulas for resistance based on displacement and speed.

Modeling and the use of iron would also help advance submarine design. Submarine technology did not stop with the Civil War and progressed with enthusiasm. Civil War submarines such as the *Hunley* were wrought-iron boiler plate (in the case of the *Hunley*, an old boiler), but as Jules Verne envisioned, steel would become the skin of choice by the 1880s. The Civil War submarines had pioneered the use of a screw propeller for propulsion, but these screw propellers were hand-cranked. Steam engine screw propellers had been used on the low-water-line ironclads. Still, steam engine submarines could not submerge. Some inventors used a combination of steam power for surface running and electrical battery power for underwater. John Holland in America had designed a submarine closer to that of Jules Verne's fiction in 1876. Holland's submarine used compressed air for breathing and for ballast driving-control tanks. While using electricity, Holland wanted to eliminate steam power in order to free the mobility of the submarine. He experimented with an internal combustion engine but never got it successfully operating. The British submarine the *Resurgam* used this combination of steam and electricity in 1879, as did the French *Goubet II*. The *Resurgam* sank off the Welsh coast, killing three crewmen. The situation improved in 1885 with the invention by Gottlieb Daimler of the internal combustion engine. This diesel and battery combination proved successful for smooth operation. As might be expected, the Germans took the lead with the Nordenfelt submarine in 1890. Holland produced a similar model in 1896 known as the *Plunger*. Holland used electrical batteries, as Jules Verne had in his fictional submarine, for submerged propulsion and steam for surface running. Holland had continued his work with the introduction of the *Holland*, which used a gasoline engine for surface running and an electric motor for submerged running. Holland also went on to be the first to apply Jules Verne's double-hulled design. Some of Holland's early models can still be seen in Paterson, New Jersey. Progress was so rapid that by the beginning of World War I, there were around 400 submarines in service across 16 navies. The Germans had the only open-sea submarines because of their earlier progress with diesel engines. Most non-German submarines were required to hug the shoreline.

8

Aluminum — Victorian Gold

Is it possible that Charles Hall in Ohio and Paul Heroult in Paris made simultaneous yet independent discoveries of how to refine aluminum? History has given them a draw, allowing America and France to choose. There is even a claim of an earlier process by American metallurgist William Frishmuth. Again, however, we see a focused stream of technology coming to an eventual point of birth. The battle is a nonferrous mirror image of the Kelly-Bessemer ferrous story. Hall, like Kelly, was a trained chemist and experimenter. Heroult, like Bessemer, was a multidisciplined inventor. Heroult is probably best known for his invention of the electric steel furnace, which was accepted throughout the world and even today produces the majority of steel produced in the world. History, like sports, does not like draws. Certainly, Heroult and Hall had an equal start, being born in 1863, and both applied for a patent in 1886. Furthermore, both had some college chemistry, although Hall graduated in chemistry and Heroult flunked out. Both worked many experiments in isolated makeshift laboratories. Both were driven by the huge potential financial rewards of the process. Both claim to have been lovers of Jules Verne's novels. The real story may be how many others were within months of the same discovery. We know that within a year, Austrian chemist Karl Joseph Bayer reported an improved version of the process. Bayer was the (Robert) Mushet of the aluminum refining process. Bayer made the Hall-Heroult process a commercial success. The story of the "simultaneous" discovery of aluminum refining is a clear example of the streams of creativity theory. The headwaters of the refining of aluminum goes back to Davy, Oersted, Wöhler, and Sainte-Claire Deville of the early 19th century, although the story is an old one.

The story of aluminum might have a 2,000-year gap in it. There exists an amazing story attributed to historian Pliny the Elder.[1] A stranger was said to have given a unique metal cup to the Roman Emperor Tiberius. The cup was shiny like silver but was said to have been made by a secret process from clay. Tiberius feared that such a metal would deflate the value of his gold treasury. Tiberius's solution was to behead the inventor and destroy his workshop. There is one piece of evidence that suggests the some ancients knew how to produce aluminum. An archeological find in China of a third-century emperor Chou-Chou revealed an aluminum ornament that cannot be accounted for.[2] Other than these strange stories, aluminum is clearly a Victorian metal, being first suggested as an element in 1808 (although some even debate this, attributing it to Lavoisier in 1787) by Sir Humphrey Davy. Davy was a brilliant chemist, having discovered potassium and sodium a few years earlier. Davy identified alum as an oxide of a metal, but he never successfully separated the pure metal. Davy's work with aluminum, his concepts, experiments, and writing, were spreading throughout Europe and America. Davy gave it a Latin name, aluminium, which he later changed to aluminum. The idea of a metal in clay stirred imaginations. Clay is the most common mineral on earth (aluminum is the third-most abundant), and elemental aluminum is the most abundant metal on earth. The concurrent developments of aluminum refining continued in the 1820s with claims of separation of the metal.

Generally, Danish chemist Hans Christian Oersted is believed to have first obtained a metallic specimen of aluminum in 1825. Oersted himself was studying electrolysis and put little significance on the byproduct of aluminum. Oersted's experiments were published throughout Europe, but two years later German chemist Friedrich Wöhler claimed the honor. The price of those early specimens was estimated at $545 a pound. The amount produced was less than a fraction of an ounce. Both Oersted and Wöhler used chemical reactions between potassium and aluminum chloride. The amounts produced were a few grains no bigger than a pinhead. Neither of the processes could be considered economical. Henri-Étienne Sainte-Claire Deville made the commercial breakthrough in 1854 with a sodium reduction process. The process was complex and required the production of several intermediate chemicals, such as sodium and anhydrous aluminum chloride. The Deville process had another advantage in that a French geologist had discovered a clay very high in aluminum content near the village of Les Baux. This is where aluminum oxide commercial deposits derived their name — bauxite. The process was good enough to bring down the price to about $16 per pound. Deville's major accomplishment may be his publication of a book, *Aluminum, Its Properties, Its Production and Applications.* The work would be a starting point for Hall and Heroult as well as Jules Verne in his later novels using the metal. Deville's book would form the basis for many predictions about aluminum made by Jules Verne. In his 1865 novel, *From the Earth to the Moon,* Verne correctly predicts its use as a space capsule material:

> As you know, in 1854 a famous French chemist, Henri Sainte-Claire Deville, succeeded in producing aluminum in a compact mass. Now, this precious

metal has the whiteness of silver, the indestructibility of gold, the toughness of iron, fusibility of copper, lightweight of glass. It's easily worked, it's plentiful in nature — since aluminia constitutes the base of most rocks— its three times lighter than iron and it seems to have been created for one ultimate purpose: to give us the material for our projectile.

Verne was a major supporter of aluminum, based on Deville's own suggestions. Verne went on to suggest the use of aluminum in buildings, planes, and boats.

Amazingly, the first of the great authors to predict the future of aluminum was not Verne but Charles Dickens. Dickens wrote the following in 1857:

Within the course of the last two years ... a treasure has been divined, unearthed and brought to light ... what do you think of a metal as white as silver, as unalterable as gold, as easily melted as copper, as tough as iron, which is malleable, ductile and with the singular quality of being lighter that glass? Such a metal does exist and that in considerable quantities on the surface of the globe. The advantages to be derived from a metal endowed with such qualities are easy to be understood. Its future place as a raw material in all sorts of industrial applications is undoubted, and we may expect soon to see it, in some shape or other, in the hands of the civilized world at large.

Dickens's excitement came from the Paris Exposition of 1855, which was aluminum's coming-out party. Deville's book had also awoken all of Europe to the wonders of aluminum.

Deville's work and exhibit at the Paris Exposition in 1855 stirred the imaginations of many. Deville heralded (as Pliny did) the metal as silver from clay. Furthermore, the exhibit put the aluminum bar with the Crown Jewels. We have already seen that the metal mesmerized Emperor Napoleon III. Napoleon III probably went even further than Verne's fictional uses in applying aluminum to armor as well as jewelry. Napoleon III's support brought the needed investment for Deville to open the first commercial aluminum plant near Paris. Deville was able to meet the project demands of Napoleon III, but his process was limited and the demand minor. Financially, Deville continued to struggle with his chemical reduction process. Still, as limited as Deville's production was, it was more than enough to meet the demand. In the end, demand proved the bigger problem in the 1850s. In 1859, Deville complained, "Nothing is more difficult to admit to the customs of life and introduce into the habits of men a new material, however great may be its utility." For almost two decades the production and application of aluminum stalled, with the exception of the great science fiction writers. In 1879, *Scientific American* reported on the lack of progress: "There are several reasons why the metal is shown so little favor.... First of all there is the price; then the methods of working it are not everywhere known; and further, no one knows how to cast it." In 1879, Ferdinand von Zeppelin designed his first dirigible aircraft but had to wait another 20 years until enough aluminum was available.

Not everyone was asleep during the 1870s; a pure Victorian engineer was learning how to make and cast aluminum. William Frishmuth improved on the chemical reduction method and was outproducing Deville. Perhaps even more important, Frishmuth was working and casting aluminum. At the 1876 Centennial Exhibition, Frishmuth exhibited an elaborate surveyor's transit made out of aluminum. He exemplifies the great Victorian metallurgists and the Victorian network of science. Frishmuth was born in Germany in 1830. He studied a year at Saxe-Weimar under Friedrich Wöhler, who had earlier isolated aluminum. Frishmuth worked as a practical chemist in the Caribbean and South America before finally settling in Philadelphia in 1855. He served as a special agent during the Civil War. He became an active researcher in the electroplating of metals in the early 1870s, which educated him in the electrochemistry needed to produce aluminum. In 1883, he patented a new and improved chemical reduction of aluminum. Not only could Frishmuth outproduce the Deville process, he could do it much more cheaply by eliminating the need for expensive solid sodium. Instead of solid sodium, Frishmuth generated sodium vapor using cheap sodium carbonate and a reducing agent. This process also greatly reduced the wear on the equipment. With a cheaper process and the ability to produce commercial quantities, he constructed the first aluminum foundry in Philadelphia. He went on to develop new casting techniques to better use aluminum. Frishmuth's process might have even became a commercial success had a cheap source of energy been available. History has forgotten most of Frishmuth's advances in aluminum production with one exception — the aluminum cap on the Washington Monument.

The Washington Monument had been started in 1848 to honor the first president. Construction problems and the Civil War halted progress for years. It wasn't until 1884 that the final pieces were to come together. Colonel Thomas Lincoln Casey, chief engineer, wanted a metallic apex on top of the monument to act as a lighting rod. When Casey contacted Frishmuth, several metals, such as copper, bronze, and brass, were suggested. It was Frishmuth who suggested aluminum. Frishmuth did reserve the right to cast it out of bronze if he could not cast the pyramid apex. The six-pound aluminum casting would have been the largest ever. Frishmuth had developed the necessary expertise to produce such a casting and was successful. On November 12, 1884, Frishmuth sent the following letter to Casey: "After hard work and disappointments, I have just cast a perfect pyramide of pure aluminum made of South Carolina Corundum. Great honor to you, the Monument and whole people of North America & a little myself lent.

Cost of pyramide more than calculated. When do you want the pyramide." The final cost was $255. The casting was completed at Frishmuth's Philadelphia foundry in early November 1884, but he delayed shipment to publicize it. He displayed it at Tiffany's jewelry store in New York. It was mounted at the monument on December 6, 1884, and the event was on the front page of *Harper's Magazine.*

Frishmuth's six-pound casting in 1884 added to his total production for that year of 112 pounds, considered an amazing record. Frishmuth's Rush and Amber Street foundry is commemorated with a historical landmark. The cast-aluminum plaque reads, "Colonel Frishmuth's Foundry.... The site of the first commercial aluminum reduction facility in the United States of America and the only producer of aluminum from its ore until the late 1880s." On November 12, 1984, the foundry cast an exact replica of the six-pound apex, and it was again displayed at Tiffany's. Frishmuth's reduction process had probably brought the cost down to under $12 dollars a pound, but future electrolytic processes would bring this to under $2 a pound. It might have been possible for Frishmuth to even lower the costs of his chemical reduction process since he was using crystals of corundum (sapphires and rubies) instead of high-aluminum clays such as bauxite. The rapid success of the electrolytic process, however, left little economic motivation for him to improve on his process. Frishmuth's process was lost to most mainstream history books, and he personally fared even worse. He did receive 12 patents, including a portable electric light and several types of batteries. In this respect, Frishmuth belongs in the Victorian hall of metallurgy and engineering as a creative genius, like so many of his time. Still, his aluminum business fell on hard times with the low-cost electrolytic processes of Hall and Heroult. By 1893, he was overwhelmed by business problems and legal fights. On August 1, 1893, he was found dead from a self-inflicted gunshot wound. Frishmuth's achievements with aluminum preceded even some of Verne's greatest fictional applications. Frishmuth, the Victorian dreamer, had created a market for aluminum and developed application techniques such as casting. Maybe just as important, Frishmuth created the whole field of aluminum metallurgy. He pioneered aluminum bronze, a strong, lightweight alloy for shipbuilding applications. Some of his other aluminum alloys were forerunners of today's aircraft alloys. The price of aluminum, however, limited Frishmuth throughout his career.

The freeing of aluminum from its ore (aluminum oxide) is a complex metallurgical problem. Iron, for example, is relatively easy to free from its ore. The chemical reduction in the case of iron is done by carbon and heat. Chemical reduction is done by using an that replaces the metal held in the oxide with another element that wants to unite more with oxygen. Carbon therefore replaces iron in the oxide, forming carbon monoxide and freeing pure iron. This concept of chemical reduction is the same principle at work in a battery. The problem with aluminum is that aluminum oxide is one of the most stable compounds on earth. Aluminum oxide is very reluctant to give up its oxygen. Chemists such as Davy, Deville, and Wöhler had discovered chemical reduction using sodium. Frishmuth made the process more efficient by using vapor sodium and additional energy in the form of heat. Frishmuth knew of Davy's experiments in 1806 using electricity as an energy source to produce metallic sodium, but he could not find a commercial power source with enough voltage. Even Davy had to use a roomful of 2,000 batteries to produce sodium, which is relatively easy compared with aluminum. Still, aluminum oxide resisted yielding in economic quantities. Both Wöhler and Frishmuth had tried to use a more efficient electrical energy source in the reduction than heat. One of the limitations that both men faced was not knowledge, but a more powerful electric dynamo to supply the current. Had large dynamos been available earlier, the secret of aluminum smelting could have been discovered before the births of Hall and Heroult (1863). Even today electrical power remains the limitation to the location of aluminum smelting plants. This is why countries with cheap hydroelectric power sources are favored, and energy-short countries such as Japan are at a disadvantage. In the United States, aluminum smelting is a small industry (60 percent of aluminum comes from recycled sources), yet it makes up more than 4 percent of our power consumption. To state it another way, the amount of energy needed to smelt just one pound of aluminum would keep a 40-watt light bulb burning for 10 to 12 days.[3]

The complexity of the process is that it requires not only huge amounts of electricity but also many intermediate steps. The modern version of the Hall-Heroult process starts with bauxite, a clay that contains about 60 percent aluminum oxide (most common clay contains 45 percent). The bauxite contains a lot of water, and it must be ground and heated in rotary kilns. The powdered bauxite is then mixed with sodium hydroxide (caustic soda) to form sodium aluminate. The sodium aluminate is roasted at a high temperature to produce a concentrated aluminum hydrate and aluminia. To apply electrolysis requires a liquid or conductive bath. Aluminum fluoride is added to dissolve the alunimia particles at a low temperature. A carbon electrode is then lowered into the liquid bath and a current applied. The carbon electrode, known as an anode, is where the carbon reduces the aluminum oxide. Carbon dioxide is generated, freeing the aluminum from oxygen. The long struggle to free aluminum from its ore created major advances in battery technology and corrosion metallurgy. Actually, Wöhler in 1827 was close to reducing aluminum had he had the electrical dynamos of the late 1880s. The first dynamos capable of supplying the necessary power did not show up until Philadelphia's Centennial Exhibition, but these small dynamos were overshadowed by the 30-foot Corliss steam engine. Still, the seeds of electrical application had been sown. The ideas of Davy and Wöhler would not have to wait long as dynamo technology improved.

The German chemist Wöhler was a mentor not only for Frishmuth but also Paul Heroult. Heroult, born in 1863, was a difficult child. He disliked school, but he loved to read, his two favorite books being Jules Verne's *Mysterious Island* and Deville's *Aluminum, Its Properties and Its Production and Applications*. From his reading of Verne and Deville, Heroult developed an interest in chemistry. Heroult appears to have briefly worked with Wöhler in his short college career. Another chemistry professor seemed to have encouraged him to experiment in aluminum refining. In his 20s, his quest for an aluminum reduction process became an obsession and passion. Heroult set up a makeshift laboratory in his father's tannery barn. In 1883, he gathered capital and help from fellow students to purchase a 30-volt dynamo and try electrolysis. His earlier attempts had failed until he discovered the addition of aluminum fluoride in 1886 that would allow for the electrolytic deposit of aluminum.

At the same time, a continent away, Charles Hall discovered the same addition of fluoride and the use of electrolysis. Heroult's patent came a few months earlier in France, and Hall was shocked to find that Heroult had beaten him. Hall had never heard of Heroult. Yet they had common information sources from Wöhler's, Davy's, and Deville's writings. Both had duplicated the earlier experiments of these great chemists. Here again, we see a stream of creativity converging to a focal point of invention. Their stories are strikingly similar. Hall was also born in 1863. He pursued a career in chemistry and was encouraged by a friendship with a professor at Oberlin College in Ohio. The search for aluminum became a passion for him as well, and led to his setting up a laboratory in his father's barn. Hall, like Heroult, hit on the use of fluoride as an agent. This idea of two "independent" inventors coming upon the use of aluminum fluoride seems amazing, but the idea was far from new. Three years before Hall and Heroult applied for a patent, Charles Bradley had suggested the use of fluoride in aluminum smelting by electrolysis.[4] Bradley's patent was rejected based on prior art! The patent office cited that Davy had suggested it in the early 1800s. This suggests that the keys to the actual process were a matter of close studies of the advances of Davy and Wöhler.

Neither Hall nor Heroult had instant success with their processes. Heroult's use of French low-silica bauxite ores did not produce a pure, immediate aluminum product in the reduction process. In 1888, an Austrian chemist, Karl Bayer, invented a process to better produce the intermediate aluminia, but Heroult did not apply this until the 1890s. In 1889 aluminum remained more costly than gold. In fact, the British government gave the inventor of the periodic table, Mendeleyev, a set of gold and aluminum scales as a token of recognition. In the meantime, Heroult struggled to produce low-cost aluminum. Heroult was fortunate to have investors ready to pour money into the development. The patent battles did not help. Power generation was another problem, but Heroult stayed the course. Finally, with the adoption of the Bayer aluminia process, the plant turned the corner. Still, French power costs put it at a disadvantage to American operations that had tapped the hydroelectric power of Niagara Falls. The French, however, continued to apply aluminum in new applications. In 1892 the French built the first all-aluminum boat, a 40-foot yacht. American companies copied the initiative with the production of aluminum rowboats in 1893. Heroult settled the patent disputes with Hall. Heroult went on to invent the electric steel-making furnace, which launched a new revolution. Heroult embodied the characteristics of the Victorian meta-inventors, such as Henry Bessemer. Heroult's imagination seemed unbounded. He invented a prototype of the modern helicopter known as the "phaneroptere" and a hydroplane boat.

Heroult's electric furnace offered a metallurgical tool that would be the key to the aircraft industry of the next century, and the Victorians were quick to develop that stream of creativity even before the electric furnace was perfected. A fellow French chemist, Henry Moissan, used the electric furnace in 1898 to electrothermally reduce two new metals from their ore— Niobium (Columbium) and Tantalum. These new metals would be key in the development of jet turbine blades and aircraft parts in the next century. The electric furnace could also be used to reduce other metal oxides, such as chromium. Again it was Moissan in 1893 who used the electric furnace to produce ferrochromium for the steel industry. The Russians used the electric furnace to reduce the cost of tungsten for making economic light bulb filaments in the early 1900s.

Hall's more efficient variation of the aluminum smelting process fared little better than Heurolt's initially. Commercial development required the input of a creative metallurgist and industrialist—Captain Alfred Hunt. Hunt had worked in the development of crucible steel methods early on. He had also formed his own company in Pittsburgh Testing, which serviced the booming steel industry. Charles Hall met Hunt in 1888, and the future of aluminum changed after that meeting. As a metallurgist (a graduate of MIT), Hunt immediately saw the potential. Hunt brought marketing, patent, and financial expertise to the partnership as well. Hunt and Hall formed Pittsburgh Reduction Company in August 1888, selling stock and ordering two giant Westinghouse dynamos rated at 1,200 amperes and 25 volts. On Thanksgiving Day 1888, they produced the first aluminum. In

1889, Pittsburgh Reduction produced 6,943 pounds of aluminum at a selling price of $2.75 a pound, and in 1892 it produced 138,307 pounds at a selling price of $.65 a pound. At that price, Hunt could market aluminum worldwide. In an 1891 lawsuit, Hunt listed the following applications of aluminum:

1. Replacement of brass metalwork
2. Replacement of nickel and German silver
3. Builder's hardware
4. Aluminum castings for machinery (the Wright brothers' plane would use an aluminum crankshaft case)
5. For the production of steel castings and cast-iron castings (this would be its main tonnage use for years)
6. As machinery shim
7. As aluminum foil for wall decoration
8. Plating cast iron
9. As horseshoes (still used today for race horses)
10. Building sections
11. Production of aluminum bronze[5] (a corrosion-resistant alloy)

Hunt addressed the market head-on since he realized that volume was the key to lower production costs. This marketing problem was one that Heroult never fully solved. Hunt created markets wherever he went. In Pittsburgh, millionaire industrialist Henry Clay Frick decorated his mansion walls with the very expensive aluminum foil in 1891. This unusual application can still be seen in the house today. Hunt was also a brilliant metallurgist who developed the use of aluminum additions to steel. The stability of aluminum oxide (aluminia) was the reason it eluded reduction for so many years, yet this very property makes it extremely useful in steel making. Molten steel by its processing, such as Bessemer air blowing, tends to be high in oxygen. This oxygen can make the steel difficult to pour and cast. The addition of aluminum allows the aluminum to gather up the oxygen, improving the casting of the steel. Aluminum is known as a deoxidizer, and even today remains a critical factor in continuously cast steel. In Hunt's day the use of aluminum in steel making made the steel industry his biggest customer. An interesting attribute of aluminum's strong affinity for oxygen is its corrosion resistance. Raw metallic aluminum will quickly form an oxide skin with exposure to air. Unlike metals such as iron, which form a crusty oxide (rust), aluminum forms a tight, thin oxide surface that protects the metal underneath. This thin oxide surface appears as metal to the eye and can be polished to a luster as with aluminum foil. This property inspired the first large casting of aluminum, the statue of Eros in Piccadilly Circus (1893).

Hunt, as a metallurgist, realized the amazing properties of aluminum and, like Verne and Deville, he saw endless possibilities. Aluminum reflects 80 percent of the light that hits it, making it an excellent reflector. It reflects heat also and finds applications in insulating such as firefighters' suits and space suits. Using elements such as copper and magnesium as alloys, it can approach the strength of steel at a fraction of the weight. Hunt early on used this property to promote its use in bicycles, and Jules Verne saw applications in aircraft. In the 1920s several car companies, such as Pierce Arrow, promoted aluminum-body cars (one of these can be seen in the Henry Ford Museum). Today truck beds, aircraft, hiking gear, and containers use its good strength-to-weight ratio. Ferdinand Zeppelin used a unique alloy of aluminum, Duralumin, containing copper and manganese as well as aluminum, in his 1901 dirigible. Duralumin dominated the aircraft industry until 1938. Aluminum is a great electrical conductor. At about a third the weight, of copper it is often used in large power cables, which can span longer distances between towers or pylons. Aluminum is also a good heat conductor, which led Hunt to use it in the production of pots and pans. Today, this property has made it popular for car radiators and engine blocks. Only the higher cost of aluminum compared with steel has limited its use.

There is still one Victorian prediction about aluminum that has not been realized, but might well be in the future. H. G. Wells, in his novel *War of the Worlds*, described Martian aluminum production: "From sunset to the appearance of the stars this clever machine manufactured at least one hundred strips of aluminum directly from clay." Wells was predicting solar power as the electrical power for smelting the red "clay" of Mars. It is probably unlikely such an operation will be used on Mars, but the moon might see it yet this century. Moon rocks suggest aluminum concentrations as high as 15 percent or 200 tons of aluminum per hectare. Harnessing solar power on the moon for such an operation is certainly within the realm of possibility.

9

Basic Victorian Electricity

The era of electricity, while having roots in the previous century, started concurrently with the 19th century. A few years ago, the status of electrical science in the early 1880s was summarized: "That science [electrical] had not greatly advanced beyond Franklin's achievements, but in 1800 the Italian physicist Alessandro Volta (1745–1827) had announced his new source of electricity, the voltaic pile: the first wet battery. Benjamin Franklin coined the term, battery, in 1748. For the first time scientists had a source of continuous current, and they soon began to discover new phenomena." Volta reported the first design of an electrical battery to the Royal Society of London on March 20, 1800. This battery consisted of a series of zinc discs, salt-soaked pasteboard discs, and copper discs. He had used and applied what others had unknowingly observed: that two dissimilar metals create an electric current. Volta set up three of these disc piles and produced a one-volt current, "volt" being a standard measure of electric current push. The simple principle is, the more volts, the more "potential" for a current to flow. Volta's discovery was hailed immediately as a breakthrough. In 1801, the Italian Volta was invited to speak before the French National Institute. On his arrival he was received by Napoleon Bonaparte, who spent several hours with him. Napoleon awarded him the institute's gold medal, and later he received the French Legion of Honor. Volta had applied the basic scientific and metallurgical principle of electron flow caused by the inherent difference between two different metals. Any time two metals come in contact, there is a flow of electrons (current). Volta joined copper and zinc, but any number of metal combinations could have been used. Volta's principles of the early 1800s formed a base for a century of research and application.

The trail of Volta's scientific principles goes back to the beginning of written science, but in this case it is convenient to start with Luigi Galvani (1737–1798). Galvani built his laboratory at the University of Bologna. After Benjamin Franklin, Galvani was the most published experimenter of the 1700s, his most famous experiment being the twitching of frog legs hung on an iron railing and probed with copper hooks. This basic principle of two dissimilar metals creating a current (flow of electrons) is called a Galvanic cell. It is the underlying explanation of all the corrosion of metals, electroplating, and electrical reduction of metallic ores such as aluminum. When you bring two different metals into contact, you create a Galvanic cell, or couple. If the contact is tight or an electrolyte such as water is present, an electrical current is created. One of the metals becomes the "anode" and one the "cathode." The anode has what metallurgists call higher potential; that is, it wants to give up electrons. What actually happens when a metal gives up electrons is that it corrodes, creating an oxide such as rust. The other metal is known as the cathode, and it accepts electrons and therefore is protected from corrosion. This is why galvanized steel (a zinc coating) protects steel from rusting. The zinc actually corrodes to pump electrons into the steel and protect it from corroding. The zinc is the anode, and steel is the cathode in the Galvanic cell. It is also why zinc blocks are attached to steel-hulled ships to protect them from corrosion.

The idea that metals can protect metals is fundamental in design. In electroplating the principle is the same. In this type of cell a current is applied to a metallic solution such as that of copper sulfate through an iron cathode. The iron, having a lower potential, is the cathode, and copper is plated on it. It is also the secret of adding an aluminum block (anode) to a baking soda/water solution (electrolyte) then adding silverware (cathode) to replate silver. For people like Volta and Galvani, the issue was to identify the order of potential in metals, which is known as the Galvanic series. The Galvanic series tells which metal will corrode and which will be protected when the two are combined. It also predicts the strength of the current produced based on the difference between the two metals' position on the series.

Galvani had started to identify the natural order of metallic potential through his frog leg experiments. First he found that if the hook and railing were both iron, there was not an effect. He then used a copper hook and iron to create a current, which caused the frog legs to move back and forth. When he substituted silver for iron hooks, he found a more violent and long-lasting effect. Silver had higher potential than iron, and therefore a Galvanic coupling of iron and silver created more electric current than a cell of copper and iron. Galvani, however, attributed the current to "animal electric fluid." Volta went even further in his experiments, as seen in his 1800 publication: "Yes, the apparatus of which I am telling you and which will doubtless astonish you, is nothing but a collection of good conductors of different kinds arranged in a certain manner. 30, 40, 60 pieces or more of copper, or better, of silver, each laid upon a piece of tin, or better, zinc, and an equal number of layers of water, or of some other humor

*Galvanic series of metals used to predict the reaction
of two dissimilar metals in contact*
The "anodic" or corroded end
Magnesium
Zinc
Aluminum
Cadmium
Carbon steel
Cast iron
Chromium steels
Lead–tin solders
Lead
Tin
Nickel
Brasses
Copper
Bronze
Silver
Titanium
Graphite–(used in batteries)
Gold
Platinum
The "cathodic" or protected end

which is a better conductor than plain water, such as salt water, lye, etc., or pieces of cardboard, leather, etc., well soaked with these humors; such layers interposed between each couple or combination of different metals, such as alternative succession, and always in the same order, of these three kinds of conductors, that is all that constitutes my new instruments." This simple experiment of Volta at the start of the century inspired 200 years of battery development, electrical technology, electroplating, metal ore reduction, and corrosion prevention. Battery metallurgy advanced with the almost endless combinations of metals. Robert Wilhelm Bunsen introduced the familiar carbon-zinc battery in 1842. Jules Verne proposed light metal batteries such as today's lithium batteries. Verne combined sodium and mercury in a fictional forerunner of today's powerful batteries. Even more foretelling was Verne's use of electrolysis to manufacture sodium on the *Nautilus*. Reading Verne's stories offers a detailed history of Victorian battery development.

Volta's work energized the fields of electrical and metallurgical engineering. Volta's scientific accomplishments earned him the title of count.

Volta's work also points out the nature of Victorian science. The prudent path of traditional science would have suggested that the battery's efficiency be improved and its cost reduced. Victorian science, however, raced to new applications and uses of the battery versus improvement. This occurred because the Victorian metallurgists were by nature inventors and innovators, not "brick laying" scientists. The generalists of the Victorian period favored the quest for economic applications rather than specialized scientific advances. Victorians wanted to know how a finding could be used or how to use it to experiment within a new field. The Victorians favored engineering versus pure science. They aggressively searched out applications for any new scientific principle uncovered. Applying new technology in new ways is a characteristic of the meta-inventor. Volta's experiments started within six weeks of Galvani's discoveries. This period of electrical experimentation resulted in an explosion of creativity.

Volta's battery gave experimenters a continuous electric current to examine new phenomena. Metallurgic experimentation was of course an obvious area for application and experimentation. This new experimentation generated new applications in Victorian metallurgy rather then trying to improve the battery power. One of the earliest experiments was a macabre use that formed the basis of Mary Shelley's *Frankenstein* (1818). Galvanic work in the late 1790s with dissimilar metals clearly inspired Volta's battery design. Volta, however, gave scientists a continuous electrical current to perform new experiments and apply it in new ways. A nephew and collaborator of Galvani, Giovanni Aldini, reasoned that electrical current would make dead muscles move, and maybe even create life. To test his theory using Volta's battery, Aldini worked out an arrangement with local Italian authorities in 1802 to use the cadavers of executed criminals. These tests were actually well publicized in scientific journals. The following was published in *Bibliographical History of Electricity and Magnetism* on January 17, 1803: "After the body lain for an hour exposed in the cold it was handed over to the President of the London College of Surgeons who co-operated with Aldini in making numerous observations to determine the effect of galvanism with a voltaic column of one hundred and twenty copper and zinc couples." An even more graphic description of the experiment was given in the February 1803 edition of the *Annual Register*: "On the first application of the process to the face, the jaw of the deceased criminal began to quiver, and adjoining muscles were horribly contorted, and one eye actually opened. In the subsequent course of the experiment, the right hand was raised and clenched, and the legs and thighs were set in motion, and it appeared to all the bystanders that the wretched man was on the point of being restored to life."[1] These graphic and chilling experiments not only encouraged restrictive laws, but also produced a wave of interest in electricity.

Of more practical application of Volta's battery was the work of Humphrey Davy (1778–1829). Davy was one of the first metallurgists to use electric current to reduce metals from their oxide or salt compounds. We already have seen that this work was critical in the ultimate development of the Hall-Heroult aluminum reduction process. Prior to Hall and Heroult, Davy started to pass current from Volta's batteries through alkali metallic salts. These experiments resulted in the discovery and isolation of metals such as sodium, potassium, calcium, strontium, barium, and magnesium. Using electrochemistry, Davy isolated 47 elements,[2] many times the number discovered in all history prior, and 41 percent of all elements isolated to date. Davy also made one of the first applications of the Galvanic series in corrosion prevention. In 1823, the British navy was struggling with the corrosion of copper-sheathed wooden hulls. Davy was knowledgeable about the theory of Galvanic cells and attached cast-iron blocks (cathode) to the copper hull (anode), which "protected" the copper. His most important discovery may have been making a current jump between a gap of two carbon contacts. This was the first carbon arc lighting lamp, and it found new applications quickly. Davy was from the key English mining district of Cornwall. Many miners died every year due to open candle lamps causing mine gas to explode. The arc lamp proved to be a safe replacement. Davy was one of those Victorians that created another generation of new applications and experiments through his many lectures at the Royal Society and his many publications. One of his greatest discoveries may have been his pupil, Michael Faraday. He was a prolific writer, and his 1815 edition of *Elements of Chemical Philosophy* was the century's bible of electrochemistry. His final book documented his retirement at Lake Geneva — *Days of Fly Fishing.*

Another electrical current experimenter was a Danish scientist, Hans Christian Oersted (1777–1851), who found that a compass needle was deflected by an electric current. He had discovered the phenomenon of electromagnetism, which started a revolution in telegraphy. He proved what many had suspected for years. Sailors and navigators had reported the effect of lightning on their compasses. Benjamin Franklin had reported magnetized needles from discharges of static electricity. Even farmers had observed that steel kitchen knives that had been struck by lighting became magnetized. Until Oersted's work, scientists had studied magnetism and electricity as separate and distinct forces. This electromagnetic theory represents a confluence or node of creative streams. This scientific relationship would open up the fields of telegraphy and electrical power generation. Using the constant current of Volta's battery, Oersted demonstrated a relationship between electricity and magnetism. As part of the Victorian scientific movement, Oersted published his work in 1820. His original work was published in Latin, allowing it to be readily translated throughout the world. This small piece of information then got to experimenters and scientists throughout the world. This caused further experimentation by André-Marie Ampère, William Sturgeon, Baron Paul Schilling, Michael Faraday, and Joseph Henry. The explosive nature of electrical discoveries and applications in the 1800s was unequaled until the personal computer revolution. Ampère advanced Oersted's work directly. Ampère manipulated the direction of the current-carrying wire and used spirals to increase the magnetism. In particular, Ampère developed a mathematical model and formula that had been lacking before. The imagination of Ampère's writings were also significant in that he suggested sending and receiving messages via the electromagnetic effect as well as the idea of the earth's iron core creating a magnetic field. One of the first practical applications of Oersted's work was the development of the electromagnet by William Sturgeon (1783–1850) in 1825. Realizing that Oersted's experiments with currents involved creating a magnetic field, Sturgeon tried to improve the magnetic field's strength. Using a battery as a current source, Sturgeon wrapped wire carrying the current around a piece of iron and created a strong electromagnet. Sturgeon was just one of many who were inspired by Oersted's principle. To many, Oersted is the father of the telegraph, fax, and telephone.

The importance of Oersted's work can be seen in Samuel Morse's 1856 visit to Oersted's laboratory. The following is a summary of that visit by Kenneth Silverman, Morse's biographer: "Morse considered his visit to Copenhagen a pilgrimage.... The same day he visited the study of "the immortal Oersted." He sat at the table on which the Danish scientist had observed that a wire carrying a current deflects a compass needle — a discovery, he said, that 'laid the foundation of the science of electromagnetism, and without which my invention could not have been made.' He bought a bust of Oersted at the Porcelain Museum, whereby he encountered Oersted's daughter — the living likeness of the bust of her celebrated father."[3] Oersted's work formed the basis of telecommunications for two centuries, yet his name is all but unknown except in scientific circles. Oersted's electromagnet was a Meta invention like Heroult's electric furnace that spurred new streams of research and development.

In America, Joseph Henry (1797–1878) was coming to the same conclusion concerning electromagnetism. Besides his scientific work, Henry offers a unique look at Victorian streams of creativity. He had started his career reproducing Franklin's lightning experiments of the 1700s. Henry was most often the bridesmaid because he was not interested in patenting his discoveries. He might have beaten Oersted to the discovery of electromagnetism and Sturgeon to the development of the electromagnet. He probably did beat many to the use of the telegraph. While Henry lost the historical acclaim, he was the precursor of Thomas Edison in engineering applications of scientific principles. The lifting power in Europe of an electromagnet was about 11 pounds by 1830. Henry realized that lifting power was related not only to the number of wire wrapping around the iron, but the size and type of battery. Henry used an advanced battery over Volta's battery by immersing plates of copper and zinc in a jar of dilute acid. In 1831, Henry easily lifted 750 pounds with an electromagnet. Henry's success took the electromagnet out of the laboratory and into the field. In particular, Penfield Iron Works just north of Fort Ticonderoga purchased two of Henry's magnets to separate iron from iron ore. This led to the honor of having the area renamed Port Henry.

The next step in the development of electricity was the idea of reversing the process that used magnetism to produce electricity. Henry again probably had already discovered the process when Michael Faraday (1791–1867) announced his discovery to the Royal Society on November 24, 1831. Henry in later years lamented his lack of publishing and patenting: "Henry's daughter later commented that her father 'often expressed his regret that he had neglected to publish his first results.'" Within a year, Henry had taken the concept to the practical application of electric generators and dynamos. Henry, like Kelly and Frishmuth, would, however, be in the footnotes to his rivals' patent and publishing successes. Faraday's contributions to the Metallurgic Age, however, went beyond scientific theory. Faraday was an active member of science societies, and he popularized science with demonstrations and hands-on lectures. He was the "Mr. Wizard" of his day.

While Oersted conceived of the scientific principle, Henry and Faraday launched another revolution with their induction coils in 1831. By moving a magnet back and forth through a copper wire coil, Henry produced an electric current. Faraday, if not the inventor, at least improved on this new way to produce electricity, known as a dynamo or generator. A dynamo is a machine that rotates a copper coil of wire between the poles of permanent magnets. As long as the rotation or motion continues, a current can be induced in an external circuit. The electricity is known as alternating current, as opposed to the direct current of a Volta battery. Now the world had another source of current to experiment with.

In the 1830s Henry and Faraday, along with their rivals, were looking at other applications of current from Volta's batteries and electrical generators. Communication was one of those applications. The idea of Oersted that a current could move a compass needle suggested to many, such as Ampère, that a type of signal could be sent by current. The story of long-distance telegraphy starts before that of electricity. Even in the Middle Ages, Italian castles sent messages across country by tower fires. A coded system of fire on and off was used for letters. In the 1790s, Claude Chappe built a number of towers just in sight of each other (roughly 10 miles apart). Wooden arms on top of the tower signaled letters through various positions. Telescopes were required to pick up the signals. The message was sent relay style. By 1805, Claude's system had 550 towers and covered 3,100 miles. A message could be sent 500 miles in three minutes, but of course the system was weather-dependent and daylight-dependent. Still, to send a message that fast bordered on miraculous for the Victorian mind. Communication at the start of the 19th century hadn't changed since ancient times: The news of Napoleon's defeat at Waterloo took as long to reach Paris as Scipio Africanus's victory over Hannibal to reach Rome from Africa. Claude's system would have delivered those reports in less than an hour. The system became known as the semaphore. The semaphore continued to operate until the telegraph replaced it in 1855. The first application of the "Oersted effect" predicted by Ampère was in 1832. Baron Paul Schilling built a needle telegraph for Russian Czar Nicholas I based on Oersted's principle. The needle telegraph sent coded messages between St. Petersburg and his palace 20 miles away using Volta's batteries for current. The following year, another telegraph system emerged in Germany. Karl Gauss (1777–1855) and Wilhelm Weber were studying the earth's magnetic field at Gottingen. They devised a novel telegraph for sending messages between the physics laboratory and the observatory. The system caused a deflection in a magnetic needle, which then moved a mirror, sending a light-beam signal. Even more innovative was that while they initially used a Volta battery for current, they soon applied the newly invented induction electrical generator as a current source for the telegraph.

Two British inventors, William Cooke and Charles Wheatstone, improved on the needle telegraph and built one of their own in 1837. Their telegraph ran 13 miles along a railroad in 1839. Not surprisingly, around 1837, Samuel F. B. Morse had come up with another variation of the telegraph. Morse's telegraph had some unique characteristics, but it did not break any new scientific ground. Years later, in 1849, Joseph Henry would state as much in a disposition over Morse's patent. Henry denied that "Morse made a single original discovery. I have considered his merit to consist in the invention of a particular instrument and process for telegraphic purposes." Henry's statement infuriated Morse, but such a statement actually put him in good company. The telegraph represented the evolution in the applications of Oersted's basic science. Without Morse, the telegraph would have appeared by 1840 anyway; without Oersted, it would have been decades later.

Morse was clearly not the inventor of the telegraph, yet history has rewarded him. Morse, very much like Bessemer and Hall, won the publicity and commercial battle. Morse's first love, however, was art. Over his career he produced more than 300 portraits and numerous historical scenes. He became the first professor of fine arts at an American college (New York University) and founded the National Academy of Design. This background served him well when touring Europe in the 1830s; Morse drew detailed drawings of electrical devices as he traveled. These drawings allowed him to make comparisons. The diverse interests of many Victorians such as Morse seem to be part of their spark of creative thought. It was an integral part of their creative nature and allowed many engineers to "think outside the box," an ability many of today's specialists lack. When Cooke and Wheatstone announced their telegraph, Morse's telegraph was a crude apparatus. Morse's main advantage in 1837 appears to be that he was one of the first to use the words "electric" and "telegraph" together.[4] Awakened to the advance of Europe's development of the "electric telegraph" in 1837, Morse started a campaign to claim the title of discoverer while at the same time improving his equipment. In October 1837, Morse applied for a patent of his "American electro magnetic telegraph." He then turned to the chemistry department of his New York University. There he teamed up with scientific colleague Professor Leonard D. Gale. Gale pointed out that one shortcoming of Morse's device was a "one cup" battery that is a single pair of copper and zinc plates.[5] If this analysis is true, it is clear that Morse's apparatus could have sent a message about to the next room. Gale increased the battery to 40 pairs. In addition, Gale applied the earlier work of Joseph Henry, greatly increasing the windings and therefore the strength of the magnetic field. With Gale's improvements, Morse's telegraph became a practical invention capable of sending a message a third of a mile, surpassing Cooke and Wheatstone's telegraph in 1837. By November 1837, Gale applied a new "Cruikshank" battery, which held 60 copper-zinc pairs. The idea was to achieve a range of 10 miles, the distance between stations of France's optical telegraph, the semaphore.

In the meantime, Morse developed a code to allow the messages to be sent using sounds by opening and closing the electrical circuit. With the code and Gale's improvements, Morse proposed in 1838 a 50-mile test to Congress. With the enthusiastic support of the committee chairman, Francis O. J. Smith (1806–1876), Morse got the funds easily. Smith was so impressed with the Morse telegraph that he became a partner of Morse. Smith exemplified a new type of Victorian — the heroic industrial capitalist. With the government money in hand, Morse planned to go to Europe to apply for patents. Morse and Smith first went to England, but found only rejection there. Wheatstone by then had the patent and had demonstrated success in sending messages 19 miles. Still, Morse's sound transmission stood out as superior to the Wheatstone needle. Russia and France, however, again offered little hope for Morse. Both France and Russia had national projects well under way. Morse and Smith, however, continued to take every opportunity to visit and study their rivals on these sales trips. Morse and Smith returned to America convinced they could win the commercial battle.

Morse's government project would consist of the longest electric telegraph in the world, connecting Washington and Baltimore and covering 160 miles. The plan was to use copper wire encased in lead pipe buried under two feet of earth. The 160-mile objective challenged the best combination of technology at the time, but it is these very challenges that history rewarded. Taking technology to new heights was typical of Victorian hallmarks, such as the Crystal Palace, the *Great Eastern*, the Eads Bridge, the Eiffel Tower, the Rodman cannon, the *Monitor* and so many others. Morse's effort was more like the American race to the moon in that he started the project before all the technology was even fully developed. In particular, scientists such as Peter Barlow had discovered the natural law that current diminishes along a wire at the rate of the square root of its distance from the battery. To many scientists this law suggested the impossibility of long-distance telegraphy, but in the end Victorian innovators overcame it. The ability to overlook scientific impossibility is rooted in the amateurism and diverse interests of these Victorians. Experiments by several had suggested bare copper wire caused static, or interference. Morse had been working to improve wire insulation to address part of the distance issue. He finally settled on cotton-thread windings with two coats of varnish. After months of trenching and lead pipe problems, it became clear that lead pipe conduit would not do. After 10 miles, Morse halted the project for further study.

Finally, Morse decided to return to an alternative method he had proposed to Congress six years earlier. The alternative was to stretch the wire in the air using wooden poles. Morse moved to the use of chestnut poles about 30 feet high and 200 feet apart to string his wires. He still faced battery power problems. He originally had suggested relay stations connecting circuits every 20 miles, which could resolve the problem. This arrangement allowed for freshly supplied battery power every 20 miles, and this could be his fallback design. During the project a more powerful battery was invented. This Grove battery consisted of 100 paired cells using a new metal — platinum — and nitric acid. The battery was more than capable of sending current all 160 miles, but Morse still designed in some relays. Still, as construction moved eastward, Morse tested the signal along the way. Since he needed to solder the wire every two miles, these connections required testing. On the good days the project made about a mile a day. Like all great Victorian projects, the press and visitors dogged every advance and setback. Both the press and these visitors were, however, a key part of

Morse's strategy to win commercial and, ultimately, legal success. Since the telegraph followed the railroad, Morse had an unofficial "opening." On May 1, 1844, the Whig party in Baltimore nominated Henry Clay for president. When the news had reached the train stop of Annapolis Junction about 22 miles from Washington, he demonstrated the first practical application for the new communication age. As the train coming out of Baltimore stopped at Annapolis Junction, an assistant took the much-awaited news and telegraphed it to Morse in Washington. Morse spread the news an hour and half before the train arrived. The stunt was a dramatic success, and the press loved it. The recorded opening of the telegraph on May 24, 1844, was an anticlimax except as viewed in the present day.

The now-famous transmission of "what hath God wrought" actually got little attention in the press of the time. Morse was too good a showman to let this opportunity pass, so the following day Morse relayed messages from the continuing Whig Convention, to the delight of congressmen and the press. The infusion with politics inspired the vision of reporters. Even more so, the telegraph was a technical and social revolution to a waiting Victorian world. William Bryant would call Morse the man who "annihilated both space and time in the transmission of intelligence." A newspaper report at the time went even further: "Steam and electricity, with the natural impulses of a free people, have made, and are making, this country the greatest, the most original, the most wonderful the sun ever shone upon.... Those who do not mix with this movement — those who do not become part of this movement — those who do not go on with this movement — will be crushed into more impalpable powder than ever was attributed to the car of Juggernaut. Down on your knees and pray." The reporters understood more than the public realized that the world of communication would change forever. In 1846, another key telegraph line was completed between Washington and New York. The main customers were the newspapers. In 1848, the Associated Press formed to pool news groups and lower the cost of using the telegraph. In Europe, Baron Paul Julius von Reuter (Reuter News of today) started a news agency connecting major European cities in 1851. National borders caused problems, to which Reuter responded by using homing pigeons to connect the gaps. Headline news could traverse the continent in minutes by 1858. The term "wire service" still remains today for international news agencies.

Reuter developed the first transatlantic news network. The news of the peace treaty of the War of 1812 took six to eight weeks to arrive in New Orleans from London. The news of the death of Napoleon on St. Helena took two months to reach London. Most international news in the 19th century worked its way through an elaborate system of agents, shipping companies, travelers, and political networks. Lloyds of London had established an infrastructure of 6,000 "agents" to move news around the world. The public appetite in Europe for news of the Civil War allowed for a major commercial opportunity for Reuter. Reuter arranged for American telegraph messages to be taken from eastern ports to fishing ports in Canada and the North. At the these points the messages were transferred to Irish boats, which would reach Ireland in a week and then be telegraphed to London. The system was fine-tuned by the end of the Civil War. Twelve days after Lincoln's assassination, the news reached London.

Additional telegraph lines were developed quickly throughout America; Europe followed quickly with most major cities being connected by 1865. The rapid success of Morse's telegraph overshadowed another variation of Oersted's principle: that of the facsimile. A Scottish clockmaker, Alexander Bain, applied the ability to move a needle over a distance to a machine he called the "telegraph." Bain received a patent in 1843 for "signal telegraphy" using electrical currents. Bain never fully developed the facsimile, but a British inventor, Fredrick Blackwell, demonstrated it at the Great Exhibition of 1851. In 1865, a French inventor, with the support of Napoleon III, started a fax service sending facsimiles of letters between French cities. The French machine was known as the "pantelegraph." The commercial success of the telegraph ended the effort to further improve the fax machine for 100 years, although Edison did try to resurrect a version of it in the 1870s. In Edison's case, it was the invention of the telephone that buried the project. The telegraph, however, retained popularity well into the 20th century.

As early as 1852, an American financier, Cyrus W. Field, was promoting the possibility of a transatlantic cable. Field would be the spark to bring Europe and America together on what some call the Victorians' greatest achievement. A Morse biographer summarized the motivation of the 1850s: "The simultaneous movement of consolidation and expansion came together in the most complex technological feat ever attempted, and one of the great farsighted adventures in human history — laying and operation of a submarine telegraph cable across the Atlantic Ocean, electrically connecting the Old World with the New and opening the present era of global communication."[6] It is hard for us sitting in the 21st century to put this feat in perspective. The Victorian newspapers of the time loved these dreams and engineering feats. Some called the idea "The Great Work of the Age," and "a voyage more important than any in marine annals since the days of Columbus." Morse became tied up with the work of telegraph companies, patent fights and technical improvements. Cyrus Field took over the project, forcibly moving it forward. Field would be the model of future industrial capitalists of the age, acting as financier, promoter, organizer, and ultimately manager.

Morse's interest was with the technology needed. He seems to have hesitated, but probably was more overextended with continuing projects of his own. In 1852, a British company demonstrated the possibility by connecting London and Paris via a submarine cable across the English Channel. Still, the idea of dropping a cable three miles below the surface and running 2,000 miles was equivalent to aiming the Apollo 11 capsule at the moon. The very size of the dream seemed to help Field's efforts. Eventually Field's project offered Morse a new dream. Morse began to study the technical problems with renewed energy. The cable itself presented major problems, as it would require strength and flexibility. Insulation and protection from water would also be critical. Morse's technical solution was to use ⅛-inch copper wire wound in shellac-saturated linen and coated with gutta-percha (a type of rubber). Several of these copper strands could then be enclosed in a thin lead tube backed with iron wire to strengthen it. Morse continued with experiments and even asked the advice of Michael Faraday. The final design consisted of seven strands of copper wire. Each copper wire was insulated with shellac and linen built up in three layers. The seven strands were than coated with gutta-percha, producing a rope, which was in turn wrapped with tarred yarn. The final cable was flexible and light. In 1856, Cyrus Field formed the Atlantic Telegraph Company. Field brought in one of the great metallurgical engineers of the time — Peter Cooper.

In 1857, the first effort to lay the cable started with two ships from the Irish shore. Two ships were needed because there was not a single ship capable of carrying 2,500 miles of cable weighing 2,500 tons. Both Field and Morse were on board the first cable ship, with Morse doing the electrical testing. This first effort ended about 300 miles out when the cable broke. Another effort was quickly put together but also ended in failure. A ship that had been designed by destiny for this job would spearhead the third effort. The *Great Eastern* had failed as a commercial liner. Its final cruise ended with an 80-foot gash off Long Island. The gash would have sunk any other ship in the world, but the design of Brunel had demanded a double hull. The *Great Eastern* had been repaired and was offered to Field for cable operations in 1865. The cable now had the best Victorian ship technology available for the task.

The next needed improvement was the cable itself. The insulation of the cable improved significantly using pitch-soaked hemp. For additional strength, higher-quality iron was applied. The iron was galvanized as well. Galvanizing meant putting a coating of zinc on the wire to protect it (cathodic protection) from rusting. Zinc's being higher on the Galvanic series of metals than iron creates a cell. Zinc supplies electrons to protect iron from oxidizing. Here we see that the basic principles of Davy, Faraday, and Volta that had made the telegraph possible were now being applied to the metallurgical design of the cable. The final design weighed 3,575 pounds per mile and was 1.1 inches in diameter versus the old cable that weighed 2,000 pounds per mile and was 0.625 inches thick. Because of the extra hemp insulation, the thicker cable was actually lighter in the water than the old. The final results can be summarized: "Its breaking point was much higher. It was estimated that ten miles of the new cable could hang vertically in water before it would snap from its own weight. As the route between Ireland and Newfoundland was no more than two and half miles deep at any point, that would never come about."[7]

The telephone was the inevitable result of the electrocommunication stream of creativity going back to Galvani, Volta and Oersted. A biographer of Alexander Graham Bell put it this way: "The search to transmit speech electrically was part of the burst of individual inventive activity which swept 19th century Europe and America. The invention of the telephone was almost inevitable with the number of minds bent on speeding communications and with the accumulating advances in science, and in the 1870s a Scottish-born teacher of deaf-mutes, Alexander Graham Bell, decided to make his fortune by inventing an improved telegraph and came up with the telephone as well." It is not at all amazing that Elisha Gray filed for a patent a few hours after Bell in 1876. There should have been a long line. It had been 56 years since Oersted showed that a current could produce electromagnetism. It had been 44 years since Henry and Faraday had defined the application of electromagnetism. It had been more than 20 years since Morse had shown the hard engineering of long-distance communication. Still, there remained a missing piece. Robert Bruce, Bell's biographer, summarized, "Both the theory and the technology needed for long-distance transmission of speech had been at hand since the triumphs of Joseph Henry, Michael Faraday, and Samuel Morse, well before Bell's birth on March 3, 1847. But something had evidently been missing in the men or the times or both." There was a creative leap that had to be made mentally in applying it to human voice. It might well have been that the electrical paradigm was so strong as to inhibit creativity. Even Jules Verne, who also predicted many electrical devices, seems to have missed the telephone. Verne seemed to believe the fax machine would become the tool of Victorian communication.

The resistance seems to have come from a deep-seated paradigm that excluded the idea of voice transmission. Even Edison, the most creative inventor of the period, seemed to favor the development of transmitting writing over voice. The breakthrough in thought illustrates the success of amateurs in Victorian inventive genius. Alexander Bell was a speech teacher with little education in the sciences. Just as Bessemer had broken the steel paradigm and Morse, an artist, the transmission of electrical signals, Bell broke the signal paradigm, but not alone. Others were working on "harmonic" transmission. Still, the telephone did not represent a creative node or confluence like the electromagnetic applications of Oersted. C. J. Hylander put it best: "It took an artist — and a good one at that — to succeed in giving us electrical transmission of language. Morse, as a novice in the infant science of electricity, was able to get results where trained specialists had failed, hampered as they where by traditional ideas and lack of perspective in their work." Bell had come from a family where the preoccupation was with speech and its application. Bell had spent years working with the deaf, finally opening his own school for Vocal physiology in Boston.

Jules Verne had envisioned two similar inventions that were in place when Bell started his work. These were the harmonic telegraph and the autograph telegraph. A German scientist, Hermann von Helmholtz, had published experiments in 1863 of putting a tuning fork near an electromagnet and transmitting over wire to another electromagnet, causing another tuning fork to vibrate similarly. This somewhat useless invention was the inspiration, if not the key, to Bell's work. We know Bell had spent a lot of time studying the work of Helmholtz, as did Elisha Gray. Thomas Edison was also inspired and making progress on von Helmholtz's findings. Without Bell or Gray, Edison likely would have invented the telephone prior to 1880.

The Bell instrument was far from a commercial invention. The Bell invention, unlike what most of us are familiar with today, had only one part for both the receiver and transmitter. The device required you to speak into it, and then move it quickly to your ear to hear the reply. It used a diaphragm of soft iron to pick up sound vibrations as one spoke into the device. A bar magnet was placed near the vibrating diaphragm. The other end of the bar magnet was surrounded by a coil of wire. The diaphragm and bar magnet created an induced electric current in the coil, which then could be transmitted through a wire and the process reversed at the other end. Clearly, von Helmholtz's harmonic telegraph was the basic model for Bell's phone. In Bell's original system the human voice created a very weak current, limiting the transmission distance to a few miles at best. Bell and Gray, as well as others, were following each other's progress, and their designs were similar. The telegraph companies were also following the advances, and they realized the threat to the telegraph. Records show that Edison was also working on the telephone before Bell's patent in 1876, and actually notified the U.S. Patent Office of this effort.[8] There were even others in England, Germany, and America that were getting close to the telephone as well. Bell was first, but it would be Edison who made it a commercial invention.

The Bell system used the vibrating magnet surrounded by a coil to actually generate the electric current, making it inherently weak. Edison's phone was more like an electric switch that *regulated* the current, which could be of any strength and thus much longer distance transmission. Edison had designed what would become known as the carbon transmitter. Edison's design used what was known as the magneto principle, and the carbon transmitter was without a doubt a real technical breakthrough. Edison's patent of it in 1877 would launch another of those great Victorian patent battles. Bell was first, but one historian put it this way: "While there is no doubt that Bell invented the telephone, its spread across the world would have come far more slowly without the basic and revolutionary improvement that Edison now devised."[9] Edison in 1877 transmitted a voice more than 106 miles from New York to Philadelphia. Bell's partner, Thomas Watson, even suggested the technical superiority of Edison's carbon transmitter. Ultimately the legal battles were resolved, with Edison selling the rights.

The Bell system, even with its original shortcomings, did progress rapidly. Alexander Bell turned out to be a powerful speaker and salesman. He showed his invention in 1876 at the Centennial Exhibition in Philadelphia. The Exhibition spread the word of the telephone around the world. The first telephone line came into existence on April 4, 1877, connecting a shop owner to one of his employees. Banks were quick to sign on as well. By August 1877 there were 600 telephones in operation.[10] One creative use of the telephone in Europe inspired future telephony: "At the Paris Exhibition of 1881 long queues had gathered to listen to music transmitted by telephone from a mile away, and long after the exhibition was over 'theater phones' in the boulevards were linked with Paris theaters. The most highly organized and efficient 'radio phone' service was popularized not in Paris, however, but in Budapest, when regular programmes were broadcast by telephone each day from 1893 onwards. The main features of the Hungarian scheme were highly publicized in Britain by writer Arthur Mee for whom 'the pleasure phone' opened out 'vistas of infinite charm.' 'Who dares to say that in twenty years the electric miracle will not bring all corners of the earth to our fireside?'"[11]

Bell, a true Victorian inventing professional and meta-inventor, moved on to other areas. Bell worked in the area of aerodynamics, experimenting with lighter-than-air craft and hydrofoil boats. In structural design some of Bell's work foreshadowed Buckminster Fuller's geodesic domes. Bell considered his "photophone," forerunner of fiber optics, to be his greatest invention, but few others did. While Bell was the typical professional Victorian inventor, his range and diversity were equal to that of Henry Bessemer, Thomas Edison and Joseph Henry. Another one of Bell's legacies is

National Geographic, which he started after being named president of the society in 1897. Maybe more important to the world is that Bell, like his rival Edison, developed the Victorian concept of inventor. Bell's gravestone bears the name inventor. Bell's words at the Patent Congress of 1891 defined the Victorian inventor best: "The inventor is a man who looks upon the world and is not contented with things as they are. He wants to improve whatever he sees, he wants to benefit the world." He left out only one attribute of the Victorian inventor — the hope to prosper in doing so.

10

Victorian Metallurgy

Metallurgy captured the Victorian imagination. Prince Albert's love of metallurgy and mineralogy resulted in the popularity of Mineral Hall at the Great Exhibition of 1851, which inspired generations of engineers. While the applications of electricity and electrochemistry grabbed headlines with the telegraph and telephone, another more fundamental revolution was being spun off behind the scenes. The founders of that electrical revolution are familiar — Humphrey Davy and Michael Faraday. Humphrey Davy can be considered the "father of metallurgy," having discovered more than 20 new metals. Michael Faraday would go on to establish the field of metallurgy and the science of alloying. Faraday's basic work on alloying was inspired by a desire to improve tooling. As we have seen, the technical limitation to the advance of ordnance had been the lack of tooling and machine tools. Tooling was limited by the hardness of the steel employed. Even earlier in the century (1800s), Sir Marc Isambard Brunel had been inhibited in developing his hoist because of the inability to manufacture parts. Eli Whitney had faced the same roadblock with his interchangeable gun parts. Machined parts tended to be brass and bronze in the 18th century and cast iron in 19th century. Cast iron was an easy material to machine because of its graphitic structure. Wrought iron proved much more difficult. Steel, however, was as hard as the tools available. Complex threading required by inventions such as the breech-loading cannon and guns called for the very best in tooling. Part making that had held Brunel back was greatly accelerated by the invention of the steam hammer of James Nasmyth. Nasmyth's hammer allowed for forged-steel part making, such as propeller shafts.

The secret of the machine age was to be alloy steel, which goes back to the 19th-century electrochemists such Michael Faraday and Pierre Berthier. Faraday and Berthier in the 1820s developed the first iron alloys of chromium and nickel. Faraday, the son of a blacksmith, would become the premier metallurgist of the early 19th century. Faraday's stated goal was the development of alloys that could produce better cutting tools. To that end, he began a series of chemical experiments between 1819 and 1822.

These experiments were some of the earliest recorded steel-alloying trials. Faraday, assisted by James Stodart, a blacksmith and cutlery manufacturer, started to experiment with the mixing (alloying) of elements. The motivation of these steel-alloying experiments came from Faraday's visit to a Welsh copper works in 1819. Faraday was able to study copper alloying, where gold, silver, and platinum were being used to harden copper. Of course, it had been known for years that tin added to copper created a hardened tool material — bronze. Faraday became convinced that alloying might similarly harden steel, and make better cutting tools. His initial trials with noble metals, such as gold, silver, and platinum, failed to produce any hardening in steel. However, in 1820, Faraday and Stodart did publish a paper on alloying nickel and chromium with steel that held promise for tools. Faraday's experiments also led to the development of ferrochromium, which is the key raw material in alloying chrome in steel. The use of chromium (chrome) in the alloying of steel moved slowly at first, but the Victorian network of exhibitions and publications pushed its use.

Part of this slow development was deceiving. The alloying of steel was one of the few Victorian creative streams that remained secretive. This occurred for a number of reasons. Initially, steel alloying as studied by Faraday was limited to specialty crucible steel makers. The principal crucible steel makers of the first half of the 19th century were Sheffield in Britian and Krupp in Germany. Sheffield was the birthplace of specialty crucible steels, and Sheffield steels were known for their quality. The Sheffield crucible goes back to 1740 and the clockmaker Benjamin Huntsman. Sheffield steels in particular made the best cutlery available, commanding huge market premiums. Sheffield for years had guarded the very nature of crucible steel making. Methods and steel recipes were handed down from father to son. Until 1820 Sheffield steel makers held a world monopoly on quality steel, which reinforced their clandestine operations. Workers were often required to take oaths, and special steels were produced only late at night with only a handful of workers present. Napoleon, threatened by such a steel monopoly, offered rewards to any able to duplicate the Sheffield processes. Krupp in Germany used secret agents to crack the secrets of Sheffield. In any case, Krupp did not crack the crucible process until 1816, and then he could produce only limited amounts until 1820. Krupp similarly guarded his new knowledge. Krupp Steel became a Victorian black hole guarding all aspects of its steel making. Both Sheffield and Krupp seem to have been producing chrome steels on a limited basis in the 1840s. Both Krupp and Sheffield had originally gotten the idea of using chrome from Faraday's experiments.

Chrome steel was the rare exception of steel alloying that was not slowed by the steel-making secrecy. Krupp was using it in its super steel cannons in the 1850s. Chrome steel was being produced in the United States in the 1850s via the crucible steel process. It was finding application in products such as saws, jail bars and safes prior to the Civil War. In 1865, Julius Baur of New York patented chrome steel. Baur's patent was published in the September 2, 1865, *Scientific American*, which spread its use. H. A. Brustlien about the same time in France claimed to have discovered the use of chrome steels. The French chemists and metallurgists appear to have created a stream of interest in the application of chrome. Gustave Eiffel used chrome oxide to help prevent corrosion in the Statue of Liberty, which might have led to the development of chrome stainless steels by the French in 1901. In the 1870s, Eads pioneered chrome steel in structural applications such as bridgework. Eads developed the first chrome steels using the higher-volume steel puddling process versus the small-volume crucible process.

The cryptic nature of improved strength in chrome steel would not be understood until the 20th century. Faraday had discovered the use of chrome as a hardening agent in Edison's fashion of a little knowledge coupled with endless experimentation. Blacksmiths had known about carbon, the first hardening agent for steel, for centuries. As you add more carbon, heat-treating can harden steel. Heat-treating was probably a result of serendipity by an ancient smith. It calls for the smith to heat the steel to a bright red and then quench it in a coolant such as water or oil (originally smiths did this by sticking it in a live animal). Then the steel is reheated at a low temperature (tempering) to remove stress created in quenching. Chromium is one chemical element that makes it easier to heat quench, and to achieve a higher hardness than plain carbon steel. Chromium also imparts strength and wear resistance. Chrome steels would become the foundation of a group of steels known as "tool steel." These chrome steels could be hardened into cutting tools to make things like cut grooves in rifle barrels, screw threads, and threads in breech-loading mechanisms for cannons. Chromium would be the element that launched a revolution in machine tools ,increasing cutting and machining speeds.

The metallurgist that followed Faraday's pioneering work on alloy steel was Robert Mushet. Mushet deserves the title of the "father of alloy steel" because of his industrial experiments with steel alloying. Mushet is best known for his manganese additions, which made the Bessemer Process a commercial success. Manganese, however, contributes much more to the properties of steel than originally suspected. The Russians had been producing manganese early in the 19th century due to the abundance of ores in Russia. In the 1840s, Russian metallurgist P. P. Anosov studied the properties of steel with varying manganese content. These studies were published in his classic study, *On Damask Steels*. These studies guided Sheffield metallurgist Robert Hadfield to patent a "self-hardening" steel that actually increased in hardness on stress in 1883. Hadfield's alloy found applications in bearings, replacing bronze with bearing steel, in the railroad industry, such as in rail frogs. Another property of manganese additions to steel is improved machining, but this property was one of serendipity.

In the 20th century, as open-hearth steel replaced Bessemer because of its higher quality, one exception stood out — screw machine steels. Screw machining shops noted a significant drop in machining speeds and quality with the switch to open-hearth steels. Metallurgical studies showed that the high sulfur in Bessemer steels that resulted in deterioration in most steel properties combined with the high manganese of the Bessemer Process to actually improve machining. The result was a new specification requiring the sulfur and manganese additions to open-hearth steel for machining applications. The specification became known as "Bessemer Machining Steel" and is still ordered today, 50 years after the death of the Bessemer Process in the United States.

Faraday's other experimental steel alloy was nickel, which developed even more slowly. Ancient Saxon miners had known nickel, or at least its effect, because it hardened copper ores. The superstitious miners called it "Kupfer-Nickel," meaning "Old Nick's Copper," believing the devil had cast a spell. The Chinese had learned in the 17th century to add zinc to copper-nickel ores to produce a silver-colored metal known as "paktong." Paktong (Chinese for white copper) was a lot like our coin the "nickel," which is really copper-nickel. The Victorians saw it as a replacement for silver and imported it from China to use in candlesticks and tableware. Electroplating was introduced in 1844, and silver-plated copper-nickel became a less-expensive alternative to silverware. Alloys of copper, nickel, and zinc are known as German silver, nickel silver, and copper-nickel alloys. These alloys were first used in Belgium for coins in 1860.

In 1877 the first commercial nickel mines opened on the islands of New Caledonia in the South Seas. These mines allowed for the production of pure nickel and inspired more research in nickel alloying. Metallurgists and chemists at the Paris Exposition of 1878 saw the first ingots of pure nickel. The price, however, dampened enthusiasm. Nickel supplies were limited and held back interest since the profit motive was lacking. Samuel Ritchie, an Akron carriage maker and inventor, however, had taken ownership of a major deposit of nickel in Canada around 1886. The Faraday experiments had offered little insight into the use of nickel steel, and Ritchie wanted to stir up research. He wrote Krupp about using nickel experimentally, but Krupp felt the availability offered little promise for economic development. This again was a clear demonstration of economic currents advancing Victorian creativity steams. Since the economic motive was missing in the 1880s, development was limited to governments. In this case both the French and the British were looking at nickel in steel for armor plate, and both of these governments could bear the expense.

A little-known Scottish metallurgist and steel managers, James Riley of Steel Company of Scotland in Glasgow, had inspired further research. Riley had discovered some amazing results in adding nickel to steel. Nickel steel was not only stronger but also tougher than plain carbon steel. This was a somewhat surprising result since most alloys and/or heat treatments that make steel stronger usually cause a reduction in toughness. The combination of strengthening and toughening from one element was miraculous. In an 1889 paper to the Iron and Steel Institute, Riley predicted nickel's use in steel:

> For special purposes: illustrations of these may be found in all small and special type boilers, in locomotive and other fire boxes, and in the hulls of torpedo and other similar vessels where lightness and strength with non-corrodibility are of vital importance.... These metals are equally important to the shipbuilder and to the civil engineer. This is strongly brought out in considering the immense advantage to be derived from their use in large structures. Think of this for a moment in connection with the erection of the Forth Bridge, or the Eiffel Tower.
>
> If engineers of those stupendous structures had had at their disposal a metal of 40 tons strength and 28 tons elastic limit, instead of 30 tons strength and 17 tons elastic limit in one case, and say 22 tons strength and 14 to 16 tons elastic limit in the other, how many difficulties would have been reduced in magnitude as weight of materials was reduced; the Forth Bridge would have become even more light and airy, and the Tower more net-like and graceful than they are present.
>
> Then in [to] regards the requirements of the military engineer, I am inclined to state firmly that there has not yet been placed at his disposal materials so well adapted to his purpose — whether of armour, or of armament — as those I have now brought under your notice.
>
> In what may be called their natural condition these alloys have many properties which will commend them for these purposes, and when the best method of treatment, by hardening or tempering, has been arrived at, I believe that their qualities for armour will be unsurpassed.

Riley's paper awoke the world to nickel in steel, and those most interested were in the military. The British, French and German navies began tests in 1889. The American navy tests showed that nickel steel could stand up against the best armour-piercing shells. They started to test even more applications of nickel steels, including cannon steels and propeller shafts in the battleships *Brooklyn* and *Iowa*. Bethlehem Steel, the manufacturer of this new nickel steel, supplied some of the initial test batch for use in a propeller shaft for the American passenger liner *Paris*. As with a lot of innovations, there were concerns and rumors of problems. Bethlehem Steel demonstrated the impact strength of these nickel steel ingots by drilling holes in one of the ingots and plugging the holes with dynamite. The charges were exploded with no damage to the steel. These American trials created an inventive stream that energized nickel research on both continents. Around 1889, both the reluctant Krupp Steel and an American steel maker, Hayward Augustus Harvey, reported a breakthrough in nickel steel armor. Both developed an armor plate process that hardened the surface to resist shell penetration while having a tough and flexible backing to absorb the impact. Krupp Steel reported that it had "produced a metal hard outside and resilient inside. At a strike every other plate had been rendered obsolete."[1] The American-made armor got similar reviews from the *New York Times*: "[T]he test showed this to be the most wonderful armor plate ever made." Generally Harvey won the distinction of inventor with the production of armor plate being referred to as "Harveyized" in America and "Harveyed" in Europe. The great French ordnance manufacturer Scheinder-Creusot became the premier producer of nickel-alloy armor. In the 1890s the Russian Naval Department offered a gold medal to inspire competition in the manufacture of armor superior to the French armor. The Russian metallurgist A. A. Rzheshotarsky studied the "Harvey process" and improved on it. Rzheshotarsky probably copied liberally from Harvey but lengthened the process, thereby deepening the hardened surface.

The improved armor spurred cannon development. The cannon king, Alfred Krupp, recognized competitive nickel-steel armor and the technological challenge of Nobel's new gunpowder. Krupp tests using the new powder to fire armor-piercing shells at nickel steel resulted in the shattering of both bronze and cast-steel cannon. Krupp metallurgists began work on a new steel for their cannons concurrently with their armor development. The metallurgical wizards of Krupp developed a unique steel, superior to any in the world. This steel contained about 4 percent nickel and 2 percent chromium, producing a combination of strength and toughness never seen before. This cannon-steel composition would be the standard for the next 60 years and would be the basis of Krupp super cannons of the next two wars. In addition, around 1892, Krupp invented an oil-hardened chrome steel for armor-piercing projectiles. Krupp had bested the world's ordnance makers, and Krupp himself would hail its grand victory at the Chicago World's Fair of 1893. *Scientific American* also hailed Krupp's victory. Jules Verne was horrified, having predicted that Krupp superiority would lead to German militarism (*The Begum's Fortune*, 1879), which would indeed be the future of the world's next sixty years. Krupp guns and money would lead to the evil dreams of world dominance by Hitler.

Nickel and nickel steel continued to make the Victorian headlines. The use of nickel steel allowed for stronger boilerplate and increasing high-pressure steam applications. Locomotive explosions plagued the railroads for most of the 1800s, but nickel steel helped eliminate the problem. A small amount of nickel causes a major increase in toughness. A famous Midwest railroad was even named "The Nickel Plate Railroad." The Chicago World's Fair drew extra attention and awe because the Ferris wheel's shaft was of nickel steel. In 1898 Whitney Manufacturing started using nickel steel in bicycle chains. Nickel steel became a forerunner of today's "high-strength, low-alloy" steels. The improved strength reduced section size and weight. In 1903, the engine of the Wright brothers' first plane had nickel-steel parts. In 1909 the Queensboro Bridge was the first to use nickel steel in bridge construction. With more than 65 percent of the nickel being mined in New Caledonia, major financial struggles erupted, with the Rothschilds of France and Krupp of Germany fighting for control. The threat to the United States aroused two great Carnegie steel men — Charles Schwab (amateur chemist and president of Carnegie steel) and Ambrose Monell (chief metallurgist at Carnegie Steel). In 1902 these men helped form International Nickel, which had large reserves of nickel in Canada. Monell, as part-owner of International Nickel, created a number of new applications for nickel in the first decade of the 20th century. One of these new alloys was a nickel-copper alloy to compete with nickel steel. This series of nickel-copper alloys contained roughly 70 percent nickel and 30 percent copper and were called "monels." The monels were corrosion-free with the strength of steel, leading to their use in shafts and propellers for ships. In 1909 the battleship *North Dakota* was built with monel propellers. In 1910, because of monel's ability to transfer impact energy, it was being cast into golf club heads. Research on monel nickel-copper alloys would lead to nickel alloys for the development of jet engines.

Another metallic additive, or alloy, of steel that Victorian metallurgists exploited was manganese. We have already seen the importance of manganese in making the Bessemer process a commercial success by Mushet. The metallurgy of manganese and probably Mushet's inspiration come from the early work of Russian chemists and metallurgists. Still, the real breakthrough of the use of manganese as an alloying element came from another British metallurgist, Robert Hadfield. In 1878, Hadfield produced a 13 percent manganese steel. This strange alloy steel was at first a curiosity, but in 1883, Hadfield patented the new alloy steel known even today as Hatfield's alloy. This new steel alloy had exceptional wear properties, but even more amazing was that under loading, the alloy actually got harder and wore better with use. This strange property became known as "work hardening." One of its earliest applications was in safes and locks because of its resistance to force. Thieves would find safe drilling difficult as it hardened with drilling. Early construction and earth-moving equipment represented another use. It soon found military applications for helmets and rifle barrels.

Another outstanding application was its use in high-wear railroad tracks. Another curious characteristic of Hadfield steels is that they are non-magnetic. This opened up some very special applications, such as some springs. The successful use of manganese inspired experimentation throughout the metallic series.

One of these experiments led to the use of tungsten, or, as it is known in most of the world, Wolfram. Actually the use of tungsten in steel goes back to British metallurgist Robert Mushet, who added it to steel in 1864. Mushet added 5 percent tungsten and maybe some vanadium to a high-carbon steel to produce what might have been the world's first tool steel. Mushet's "self-hardening steel" was a different alloy from any other known at the time. First of all, it hardened itself in air without any water or oil quench. Slow air hardening versus water and oil quenching allowed for sharper tools because quench cracking was avoided. Second of all, it could resist high heat without softening. One of the problems of hardened steel drills, cutters, and saws is that at high speeds, they become hot and soft. The friction of the work of course generates the heat, but the softening was inherent in steel. Mushet's tungsten steel could actually become glowing red with heat and not soften! This meant that you could increase your cutting or drilling speeds by 50 percent. Forty years later, Mushet's alloy would become the basis of modern "high-speed steels." Mushet's tools amazed visitors to the Paris Exhibition of 1865. Mushet moved tool-cutting speeds from 40 feet per minute in 1850 high-carbon tools to 60 feet per minute.[2] By alloying up to 8 percent tungsten, Victorian metallurgists increased speeds and metal-cutting efficiencies to seven times those of 1864. Another application for tungsten-steel was discovered in Russia. V. N. Lipin at St. Petersburg produced a tungsten steel gun barrel with high resistance to distortion from rapid fire. The Germans were quick to grasp the importance of such steel and began applying it to ordnance at the end of the century. The reason for the high hardness and heat resistance of tungsten steels is the formation microscopically of tungsten carbide particles in the steel. Tungsten carbide has a hardness approaching diamond and a ceramic-type melting point. During World War II, the Germans produced tungsten carbide armor-piercing bullets that could not be stopped by the best armor. By 1955 metallurgists applied new tungsten carbide cutting tools that increased speeds by a factor of 1,000 over 1864.

The possible use of vanadium in steel by Mushet in 1865 is not fully confirmed. Certainly vanadium would have greatly improved tool steel, and it is known that many Sheffield steel makers added pinches of vanadium secretly to their tool steels to gain competitive advantage. When the practice began is unknown. It wasn't until 1869 that vanadium was available in commercial amounts to alloy steel. The famous Damascus steel of the 1600s has been found to contain vanadium, which greatly enhanced its strength. The source of the vanadium in Damascus steel remains unknown. Vanadium is common in meteoric iron, and this might have been the source for Damascus steel makers as well as Mushet. Henry Ford actually rediscovered vanadium steel similarly in 1905. Ford, while watching motor races, observed a collision and believed the car had unusual steel in it. He had it analyzed to discover that it contained vanadium. The source of the vanadium was again unclear. Still, Ford had a Canton, Ohio, steel company start producing vanadium steel. Ford used it to produce some of the best axle steel in the world. It quickly found uses in armor, tools, and military helmets.

The Great Russian Victorian metallurgist P. P. Anosov unlocked one of history's metallurgical mysteries. For centuries the samurai steel swords were the standard of excellence. Samurai swords exhibited hard cutting surfaces as well as flexibility. Smiths the world over had tried to duplicate samurai steel through processing. By the 1840s Anosov established that the key to samurai steel was that it contained a small amount of the element molybdenum. The discovery seemed to have been ignored to some degree because of availability. Russia had the main known reserves of the 19th century and began producing molybdenum steels by 1886. These steels exhibited unusual properties of hardness and ductility. Molybdenum steels were superior to tungsten steels in many applications. Russia appeared to have the only interest in molybdenum steels until its surprising appearance in World War I by the British. Early in the war German shells caused panic as they cut through British and French tank armor like butter. The solution was a 1 percent addition of molybdenum to British armor. The Germans in World War II paid the favor back by using molybdenum steel in Krupp's cannon, known as "Big Bertha." Big Bertha hurled shells into Paris from 30 miles away.

By 1880 there were really two distinct steel industries. The tool steel industry, which maintained the crucible process, and the high-volume Bessemer steel industry of lore. The high-volume Bessemer group manufacturers represent the steel industry of Victorian history. The crucible specialty steel makers were small and sold steel by the pound while the large makers sold by the ton. Even in America, the few small crucible tool steel manufacturers, such as Jessop Steel, could be traced back to Sheffield. These small specialty steel makers maintained a low profile. They were tucked away in the hills surrounding Pittsburgh and lower New York State. The crucible steel process was replaced in 1906 by electric (Heroult-type) furnaces from France. The first of these electric furnaces imported from France for Crucible Steel of Pittsburgh can still be seen in Pittsburgh's Station Square Mall. Still, the secretive nature of the specialty steel makers continues even today.

Their melters were banded together in special guilds that exchanged recipes and techniques. Tool steel melters often were more loyal to their guilds than to the companies that actually employed them. One of these unusual guilds exists even today as Electric Furnace Guild, which is open to specialty steel furnace melters and superintendents. Managers of organizations are not allowed at meetings, nor are technical papers made public. The crucible process was by nature a high-carbon process. Even steel metallurgists identified themselves as "black iron" (tool steel) or "white iron" (low carbon). Tool steels, while highly alloyed, were also high carbon (1 percent carbon or higher). It is the high-carbon that combines with alloys such as chrome and molybdenum to form carbides in the steel. These very special alloy recipes were tightly kept by the manufacturers. Where Bessemer and high-volume producers moved toward standardized grades of steel, the specialty producers moved to eliminate any standard grades. It was only after World War II that it would be common to order tool steel by standardized codes based on composition versus trade names such as "Jessop's super drill steel" or "Hercules chisel steel." Hundreds of tool steel compositions existed among many producers. The crucible steel industry would switch to the electric furnaces of Heroult going into the 20th century, but the secretive nature of the industry changed little.

The Bessemer high-volume steel industry also began to development some new furnace technology, which improved quality. The high-volume steel industry's main product was a low-carbon to medium-carbon steel having a manganese addition from the work of Mushet. Once in a while, some chrome might be added for a special job. The Bessemer Process had one big disadvantage in that it did not remove sulfur and phosphorus. The higher-quality ores used in America and England allowed for reasonably low levels of phosphorus. Krupp in Germany had struggled with the Bessemer Process because of the high phosphorus level of its iron ores, but through a modification did succeed in using it. A young British engineer, William Siemens, had invented a new steel-making furnace and process. As might be expected, it was "invented" elsewhere by a pair of brothers (Emile and Pierre Martin) who developed a similar process in France. Both Siemens and the Martin brothers applied for a patent in 1856 (the same year as the Kelly and Bessemer patents)! The Siemens-Martin process (later known as the open-hearth process) could effectively remove impurities such as sulfur and phosphorous and was quickly seized for development by Krupp in the late 1860s. The first successful commercial open-hearth furnace appeared at the Paris Exposition of 1867. It was at this introduction that Krupp, as well as some early American pioneers in the process, took note. In 1870, Krupp started a complete conversion to the open-hearth process, significantly improving the overall quality of German steel.

Krupp had unknowingly invested in the process that would overtake Bessemer production in 1907, and by World War II would represent more than 90 percent of the world's production. The open-hearth process offered a number of inherent advantages. First, it allowed for the use of low-quality (cheaper) ores, since sulfur and phosphorus were removed in the process. The open-hearth furnace used gas heating and therefore could add cold scrap to molten iron from the blast furnace. This allowed for the recycling of steel scrap and a considerable reduction in cost. Another advantage was that the open-hearth furnaces held as much a 50 tons (although the original furnaces were only 15 tons) versus 15 tons for the Bessemer converter. The slow nature of the open-hearth process, 10 hours versus under an hour for Bessemer converters, gave it a disadvantage. Perhaps the real advantage of the open hearth was the ability to precisely control the chemical composition of the steel. In particular, the open-hearth process facilitated the addition of chrome and nickel because chemical samples could be taken readily to control the right amount. The quality of steel produced was exceptional, and Krupp optimized it in arms production. In America the early open-hearth furnaces of the 1860s were smaller than 10 tons and could not compete with the faster 15-ton Bessemer converters. Carnegie, when building his giant Bessemer Works at Braddock in 1875, had originally planned for an experimental open-hearth, but he never followed through.

The open-hearth process gave Krupp another advantage in the production of armor plate. Armor plate is particularly sensitive to even small amounts of sulfur and phosphorus. Open-hearth steel was by far a superior armor product that no Bessemer plant could equal. These low-impurity steels not only made superior armor but superior structural steel as well. Open-hearth steel could make better rails also, though the Bessemer Process held the rail market for decades. Carnegie, aware of the advantages of open-hearth, overcame it with a modern, efficient Bessemer mill at Braddock that could compete on the cost of making rails. Still, Carnegie realized the future was to be in the higher-quality open-hearth process. In 1887 Carnegie had purchased a failing Bessemer mill a few miles up the Monongahela from Braddock at Homestead. He appointed his new rising star, Charles Schwab, to be superintendent. Carnegie wanted the Homestead mill to produce world-class structural steel for the construction industry using the open-hearth process. Charles Schwab was sent to Europe to see the best of the steel mills there. He spent most of his time at Schneider Works at Le Creusot, France, and Krupp Iron Works at Essen, Germany. Schwab had worked his way up in the company from a laborer. He studied chemistry in his home lab and understood the metallurgy of the time. As a young employee, Carnegie gave him a $1,000 to expand his home lab. The great Krupp Works impressed him most, as well as the open-hearth process.

The open-hearth furnace represented a chemical process that truly manufactured steel to fit the customer. After he returned to report to Carnegie, they decided to totally remake Homestead into an open-hearth mill. Homestead by 1890 had become Carnegie's flagship mill, making structural beams, armor plate, and large forgings. It was the first successful commercial open-hearth mill. It would rule the world steel industry till after World War II.

One of the less-noted advances of Victorian metallurgy was that of electroplating. This was a direct spinoff of Volta's 1800 invention of the battery or "Volta pile." We have already seen how this source of electrical current had started an electrical revolution leading to the telegraph and telephone. Volta's battery prompted a plethora of experiments, from dead bodies to metallurgy. One of the earliest experimenters was done by an Italian chemist, Luigi V. Brugnatelli, who passed current through metallic salt solutions. Brugnatelli, a student of Volta, in 1805 invented gold electroplating. He reported the following in the Belgian Journal of Physics and Chemistry: "I have lately gilt in a complete manner of two large silver medals, by bringing them into communication by means of the steel wire, with a negative pole of a voltaic pile, and keeping them one after the other immersed in ammoniuret of gold [gold chloride solution] newly made and well saturated." Gold plating became extremely popular as decoration in the Victorian era. Gilding did most of the gold plating of the 18th century and early 19th century. Gilding consisted of using liquid metal mercury to dissolve gold (and other metals such as silver). The metal mercury/gold solution (amalgam) could than be painted on to an object. The object was then heated, driving off mercury vapor and leaving a thin coating of gold. We know the gilding process is an ancient one, going back to the alchemists. It was an old alchemist trick to dissolve some gold in mercury and then by heating cause gold to appear. One of the most famous alchemists, Isaac Newton, is known to have had excess amounts of mercury in his hair samples.[3] Gilding reached its peak in the Victorian era because of the propensity to overdecorate.

The Victorians used gilding in ever-bigger projects, such as in Cathedral domes. One of the largest gilded domes was that of St. Isaac Cathedral in St. Petersburg. St. Isaac's dome was constructed in the 1840s. Its surface was copper sheets, which were gilded. Even through mercury poisoning was known by the 1840s, the gilding of St. Isaac took 60 lives.[4] Rudyard Kipling once said he would prefer death to working in mercury mines, where miners suffered all sorts of maladies, including crumbling of the teeth. Mercury poisoning is also the source of the phase "mad as a hatter." Mercury was used in the 18th and 19th centuries to turn beaver and rabbit pelts into hat felts. Eventually the safer process of electroplating would replace gilding. The electroplating technology was slow to be fully employed. Brugnatelli's work was studied and improved on that of Michael Faraday in the 1830s using other metals, such as silver.

Silver represented the metal of decoration for the Victorians. Silver production reached exponential growth during the 19th century. In 1810 the combined value of silver and gold products reached $1.07 million; by 1899 silver alone was $26.11 million. In 1840, Henry and George Elkington applied for a patent for silver plating. They found tough opposition from the "Sheffield plate" industry. Sheffield plate was a product in which silver plate was rolled onto copper plate to produce a mechanical silver plate that looked and felt like solid silver. Sheffield plate was invented in 1742 and had mushroomed in popularity by the early Victorian era. Early Sheffield plate consisted of a copper base, but later 19th-century producers used German silver (copper plus nickel) because its silverlike color hides worn surfaces better. Electroplating greatly reduced the cost in the 1840s and further increased the popularity of silver plate.

The popularity and growth of electroplating expanded the understanding of basic metallic batteries and corrosion. One of the largest batteries ever created was the Statue of Liberty. Its wrought-iron structure and copper sheathing emulates the original galvanic experiments of Volta. It is not the green "noble" outside corrosion that was of concern, but the galvanic corrosion of the iron, which is sacrificial to the copper. Starting work in 1875, creator Frederic-Auguste Bartholdi and engineer Gustave Eiffel began on the 94-foot-high project. It took 10 years until it was received in America. Bartholdi completely assembled and disassembled it in France. They did realize that contact points of the wrought iron and copper would create rapid galvanic corrosion of the iron. Their approach was to "short-circuit" the contact points by using asbestos soaked with shellac, predicting the potential corrosion problem. They also used cloth soaked with "red lead," that is, chrome ore (chrome oxide) from Russia. Chrome oxide is an outstanding insulator, which on a microscope level is why stainless steel is corrosion-resistant. The use of red lead suggests that the French were further along on their ultimate discovery of stainless steel in 1901 than previously thought. They seemed to have overlooked the possibility of zinc galvanic protection, which was applied in the recent renovation. In any case, without the simple understanding of galvanic protection used, the statue might have failed within decades. Over the first 100 years the asbestos did become soaked with seawater, making electric contact and accelerating corrosion by the 1980s. The restoration replaced wrought iron with stainless steel, cutting down on, but not eliminating, the corrosive electric current. New plastic insulation was used in place of asbestos, and air conditioning was applied to eliminate moisture.

Another popular decoration of the period involved gold painting. Gold paint originally contained gold powder, but many merchants started to sub-stitute bronze powder. Bronze, an alloy of copper and tin, could be made to look like gold. Yet "gold" powder and paint products still commanded a high price. The powder was made by laborious hand-filing of solid bronze or possibly a brass alloy. The ability to machine-make a bronze powder offered great economic rewards. The inventor in this case was a young Henry Bessemer. Bessemer, realizing the potential economic rewards, went on to design a "self-acting" machine that could produce large quantities of pow-der in a short time. Bessemer researched old books in the British Museum to discover the details of the hand process and alloys used. From the descrip-tions Bessemer was able to design a type of grinding machine. To ensure his process's secrecy, he purchased parts from different manufacturers so that no one could reverse-engineer it. He even went so far as to build a win-dowless plant (using skylights for light) to help keep his secret for more than 40 years. Bessemer became the major world producer of bronze powder and made a sizable fortune, money that would help him fund inventive research much in the manner of Thomas Edison. Bessemer patents and creative endeavors would be rivaled only by Edison at the end of the Victorian era.

The metal that became the darling of the age was not aluminum but platinum. Spanish explorers discovered it in the Spanish colony of Colom-bia in the 1600s before it was recognized as an element. Its popularity soared in the 1780s, when a chalice of platinum was given to Pope Pius VI. Plat-inum, which like gold is inert to oxidation, found uses in laboratory ware and glass-making crucibles by the beginning of the 19th century. Platinum is the hardest of the precious metals, and its intense silver luster remains for years. It was used to produce the standard kilogram weight for the French Weights and Measures in 1795 because it could not be corrupted. Its use in jewelry began early in the Victorian era. Its scarcity, however, limited its wider use, but in 1822 alluvial deposits were found in the Ural Mountains of Russia. In 1870 Jules Verne hailed a new future using platinum plates in his fictional submarine, the *Nautilus*. In the 1870s Edison because interested in platinum as a light bulb filament because of its resistance at high tem-peratures to oxidation. In the 1880s platinum, because of its tarnishing resistance, started to replace silver. Russian jeweler Carl Fabergé used it in eggs and toys for the czars of Russia. Platinum quickly became the metal of royalty. Research in the 19th century led to the discovery of the platinum family — palladium, rhodium, osmium, iridium, and ruthenium.

11

Victorian Toolmaking

Abraham Lincoln had learned during the Civil War that the tools and toolmakers of industry limited the best technology. Tools are what empowered Victorian dreams, and the great engineering accomplishments of the century. Yet it was the great Victorian dreams that inspired the development of tools. The *Great Eastern,* for example, was inspired by James Nasmyth's 1839 invention of the steam hammer. The *Great Eastern,* which was to be a combination paddle wheel and screw-driven ship, would require the largest wrought-iron shaft ever made. Nasmyth made the *Great Eastern* happen. Its proposed 30-inch-diameter shaft was beyond the capability of any forging hammer in existence. Nasmyth's design was a monstrous machine that was highlighted at the Great Exhibition of 1851, but usually the tools of the great projects are never recognized. Nasmyth's forging hammer became a centerpiece of the exhibition and Machinery Hall. Lack of machine tools had slowed America's development in the Industrial Revolution. Marc Isambard Brunel left America in 1800 because the tools were not available to build his ship hoist. Tools were the overlooked machines behind the headlines in the early 1800s. These machine tools are often not viewed as exciting history, and historians tend to shy away from them. One of the greatest inventions of the 19th century appeared in 1800 without any fanfare. Even today, historians never mention its birth or its existence. That invention was the screw-cutting lathe. The screw-cutting lathe was the workhorse behind standardized parts, breech-loading guns, typewriters, sewing machines, locomotives, Babbage's first computer, and the auto, to name only a few. In addition, Victorian tool making led to improved measuring tools and new capabilities in machining tolerances.

We will start with the most unusual Victorian tool and the most unusual toolmaker. The great toolmaker and meta-inventor, James Nasmyth once said that a mechanic's most important tool was the pencil. Nasmyth special-ordered pencils to make his famous and award-winning engineering drawings. Thomas Edison required short pencils for his creative drawings, so much so, that he special-ordered short pencils from the

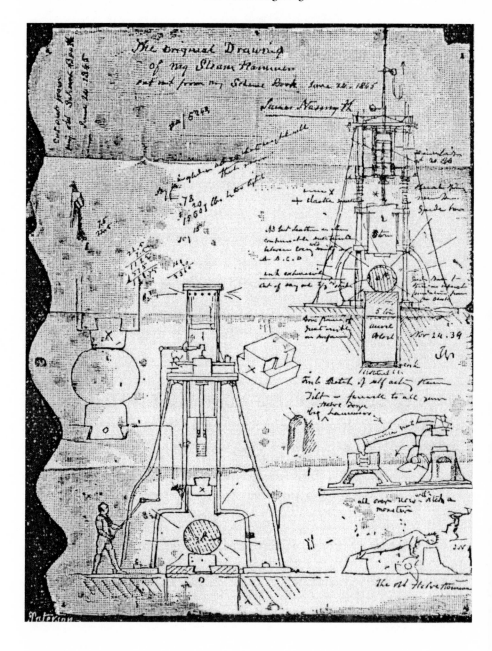

Engineering drawing of Nasmyth's steam hammer, 1839 (from *James Nasmyth Autobiography*)

factory. The following is noted on Edison's obsession with pencils: "Perhaps he settled on his ideal pencil in much the same trial and error manner in which he selected a filament for his lamp, but once he found the ideal, he stopped looking. Edison's pencils, which he ordered in lots of one thousand and always carried in his lower vest pocket, had very soft lead, were thicker than average, and were only about three inches long. Once when his order was not filled to his liking, Edison wrote to the Eagle Pencil Company that the 'last batch was too short.' He complained of the pencils: 'they twist and stick in the pocket lining.'"[1] It should not surprise the reader to find that the modern graphite pencil has two inventors who claim it–Lazare Comte of France and Josef Hardtmuth of Austria at the end of the 18th century. The man that made American pencil making profitable is best known as a writer — Henry Thoreau.

Thoreau's interest in pencil making came from his father, David Thoreau's pencil-making business. David learned graphite pencil making from a frustrated amateur chemist, Joseph Dixon. David had also invested in a deposit of New England graphite (plumbago), and combined his knowledge and resources to start a pencil-making business in 1823. The business prospered even though the quality of all American pencils was considered lower than those of Europe. Artists and engineers called American pencils greasy and gritty compared with those of Conte of France and Faber of Germany. In particular, German pencils flooded the American market. A.W. Faber was using the Conte process but had developed a unique chemical blend. The pencil makers of Europe were secretive in their methods, but the science was not. The science of pencil making, like most Victorian inventions, resulted from a stream of scientific creativity that could be tapped into with some research.

And that stream of knowledge is where the young Henry Thoreau started when he came to his father's firm. Thoreau went to Harvard's library to begin his search. It was at the library that Thoreau made his first important discovery. The Germans were using a mixture of Bavarian clay and graphite. One formula from encyclopedias of the time called for "one part plumbago with 3 of clay, and some cows hair." This was different from the typical American formula using graphite, bayberry wax and glue, which was heated and poured into pencil grooves. The Bavarian clay was not an unknown commodity in America; it was being used in the making of crucibles for steel making and in glass making. Even in Thoreau's home state of Massachusetts, two companies— New England Glass (today known as Libbey Glass) and Phoenix Crucible Company — were using Bavarian clay in their operations. Thoreau was therefore able to obtain some sample Bavarian clay to experiment with. The Bavarian clay did produce a harder and darker writing pencil, but it was still gritty. The problem was in the coarseness of the graphite. Thoreau then designed and built a special grinding mill for the factory. His final pencil was hailed: "The result of these new developments was the first American pencil that was equal of those produced in Germany."[2] By 1838 Thoreau had significantly improved the quality of the American pencil, but pencil making remained a very labor-intensive process. Unfortunately, the experiments with graphite appear to have damaged his lungs, which caused his death in 1862.

The real breakthrough of pencil making would come from a friend of the Thoreau's—Joseph Dixon. Dixon had struggled many years in the unprofitable business of pencil making, but in the 1850s he entered the profitable business of crucible making. Dixon produced graphite crucibles for the new and growing crucible steel industry of the 1850s. In 1866 Dixon patented the first automated pencil-making machine. This machine could produce up to 120 pencils per minute — a huge productivity jump over hand manufacture. Dixon started to gain a large market share based on his equipment and favorable tariff laws. The market in the 1870s was more than $20 million a year in the United States.[3] The tariffs effectively blocked the German competition except for the highest-quality artist market. This forced a number of German pencil makers to locate in America. Faber became one of those German transplants. Still, Dixon's Jersey City plant was the "birthplace of the world's first mass produced pencils." This American pencil-making machine was so efficient that is was said "the machine separates and shapes probably fifty pencils while the foreign maker is shaping one, and requires nobody to complete its work." The machinery helped to offset the very high cost of American labor in relation to German labor. This is similar to Carnegie's use of technology to overcome foreign advantages in labor costs. Certainly again we see a lesson for today. Tariffs were used initially to offset the foreign advantage, but it was creative technology that won over, even with extremely high American labor costs.

Many histories of toolmaking and machine tools start with the American Eli Whitney. Whitney became famous with his demonstration of interchangeable musket parts to Thomas Jefferson. Muskets were, prior to the 19th century, an individual piece of craftsmanship. Each gun had unique handcrafted parts. Whitney's famous proposal was to manufacture interchangeable parts for muskets. Using 10 muskets, he disassembled them, mixed the parts, and reassembled them randomly, for Thomas Jefferson. He won a government contract and Jefferson's praise. Newspapers heralded Whitney's genius and Jefferson's praise. His Mill Rock armory near New Haven, Connecticut, became a tourist attraction. It was always known that Whitney had trouble delivering on the contract, but in the 1960s, historians challenged Whitney's claim of interchangeable parts.[4] These more recent historians seem to have expected too much from Whitney's early efforts.

Whitney did indeed try to standardize parts, but he had to purchase some of the parts, creating problems. The findings that Whitney probably got his standardized machines from French sources should not be surprising. As we have seen, rarely is one person responsible for the fruition of an idea. There is no doubt that Whitney's operation was not fully integrated manufacturing, but it was a brilliantly run operation. These questioning historians miss the essence of standardization in manufacturing. The one objection, that Whitney used hand-filing to standardize his purchased parts, actually proves Whitney's contribution. Whitney did have his workers hand-file parts, but they used tempered steel patterns (jigs) to do this filing. The use of tempered steel patterns in the early 1800s represents a huge step forward in standardized parts and interchangeable parts. Whitney's tempered steel pattern is the real meta invention. Interchangeable parts were not revolutionary, but the tools to make it happen were. The use of standardized patterns or jigs allowed for the use of unskilled labor, a touchstone of standardized manufacturing. Whitney by the 1820s had developed a set of standardized inspection gauges, which allowed standardization throughout the various government arms manufacturers. The application of Whitney's patterns and inspection gauges led directly to the mass-produced, regular-issue musket — the 1842 U.S. Model.

Whitney's legacy was a passion of American armories for uniformity and standardization. Historians underestimate the contribution of standardization and measurement, but both are meta-inventions. Recent studies of sunken ships of the Spanish Armada suggest that their defeat was more related to standardization than to the brilliance of English tactics. The Spanish ships were a mass of different calibers and sizes of cannons each requiring special size cannonballs. Cannons seemed to be jammed and unused due to the confusion produced by lack of standardized ordnance in the heat of battle. Roswell Lee, a protégé of Whitney, would carry the message of uniformity and standardization throughout the arms suppliers. Lee was manager of the Springfield armory from 1815 to 1833. He created a type of national armory at Springfield. Springfield had also pioneered the standardized U.S. Model 1842 musket. The success of Springfield came from a realization by Lee that the advance of mechanical arts depended on information exchange and assimilation rather than in-house innovations. This open approach was more radical than standardization. The European approach to arms manufacture had always been secretive and extremely competitive. Lee opened the doors of Springfield to visitors and other machinists. In exchange, his machinists were welcome at other armories. This variation fueled a successful Victorian exchange of scientific knowledge that caused explosive creativity.

Historian Merritt Roe Smith summarized the impact of Lee:

At least three major streams of inventive activity converged during the 1820s and 1830s to produce the mechanical synthesis known as the American System. Two centers of this activity — Middletown, Connecticut and Springfield, Massachusetts — were located along the Connecticut River within 40 miles of each other and shared a common heritage going back to the earliest Puritan settlements of the 17th century. The third center stood near the junction of the Potomac and Shenandoah rivers at Harpers Ferry, Virginia and reflected mores and traditions quite different from those of New England. Yet, despite these cultural contrasts, all three communities developed exciting new technologies, built on one another's advances, and eventually succeeded in manufacturing firearms with interchangeable parts. Together they exercised a cumulative impact on American metalworking and shop practice that prevails to this very day.[5]

This practice of open infrastructure for development can be extremely hard to fully appreciate in a competitive environment. The greatest achievements and meta-inventions emerged from this open infrastructure. The Victorians fostered an extremely competitive environment, but they achieved advantage with cooperation. They tried to standardize practices and materials. This type of cooperative advantage offers many lessons for today's manufacturers. Springfield was also a consistent exhibitor at world's fairs, often bringing in international advances. At the Great Exhibition of 1851, three American ordnance manufacturers won gold medals for their machinery.

Those medals would soon be put to use in the Mexican-American war and the Civil War. Ordnance had captured the imagination of the Victorians since the Napoleonic wars. The armories were the center of technology and management improvement practices. Prior to the Civil War, most of the machining technology in America was developed at the nation's armories. In addition almost all toolmaking machines and tools were imported from England and France. During the 1850s, the military achieved a revolution in manufacturing, machining, and tooling that the war ensured would remain dominant. West Point, the military academy, became one of the nation's first engineering schools. The military academies trained officers in road building, basic chemistry, and mathematical calculations for cannons, physics and other applied sciences. One of the most brilliant of these "technical" officers was Captain Dahlgren. Dahlgren had an American mother, but his father was Swedish and from a long line of scientists and chemists. After some sea duty, the young Dahlgren volunteered for the study of weapons in the navy's ordnance department. In 1847, Captain Dahlgren was assigned to the Washington Naval Yard to supervise the manufacture of some new rockets. He was soon given additional duties of ordnance testing. In the early 1850s, the Naval Yard moved more into production of ordnance as well as testing. During this period Dahlgren studied the pressure on the cannon's breech and barrel. He invented his soda-bottle cannon in

1853, and it was hailed around the world for its scientific design. The Dahlgren was an 11-inch smoothbore, one of the largest cannons prior to the Civil War. The *Monitor* used two Dahlgren guns. Officers such as Dahlgren perfected machining and casting practices. The consumer market had done little to advance machining and casting. In fact, even Cleveland and Pittsburgh had only a handful of foundries prior to the Civil War. At the start of the Civil War, Dahlgren took over operation of the Naval Yard, and turned it into an efficient factory. Applying the management practices of Whitney and Lee, the Washington Naval Yard kept up with the enormous needs of the Union, slowed only by iron shortages. He was one of Lincoln's favorite artillery officers, and became known for the "champagne trials." These champagne trials became a highlight for Washington society and the military, as well as for Southern spies.

The army found its own ordnance genius—Thomas Rodman. Rodman's expertise was in the science of metal casting and metallurgy. Rodman had tested the earliest of the American Columbiads used during the War of 1812. Rodman had developed a special casting method for cannon's in 1844. The method chilled the inside of the barrel to produce a hard, wear-resistant surface. This was the method highlighted by Jules Verne in his fictional moon-shot cannon. Verne got his idea from a March 30, 1861, article in *Harper's Weekly,* which hailed the casting of the world's largest gun in Pittsburgh. The gun was the 15-inch Rodman cast at Fort Pitt Foundry. The design had the unbelievable range of 8,000 yards, which Rodman improved to four miles by the end of the war. The Confederates were able to copy the design, but not the casting method limiting their copy to a range of fewer than 2,000 yards. Rodman improved to a 20-inch gun by war's end. The Rodmans saw little use in the war, but continued to be part of the military inventory until 1918. Their size and 20-ton weight limited them to coastal defenses and fortifications. The Rodman "chilled" casting method did, however, find many applications in cast-iron products.

Whitney started a tradition in ordnance manufacture that would continue throughout the Victorian era. Another toolmaker and manager that added to both the science of machines and men was Captain Henry Metcalfe. Captain Metcalfe, who graduated from West Point in 1885, studied manufacturing processes like Darwin studied natural sciences. He recorded his observations and looked for themes and areas of improvement. From his studies he developed managerial concepts for improved productivity. He gained experience at many of the nation's armories, including Watervliet, New York. Metcalfe was particularly interested in the interaction between supervisor and machine/worker. Many of his concepts predated the principles of scientific management set forth 10 years later by Fredrick Taylor. In 1885 Metcalfe published *The Cost of Manufactures and the Administration of Workshops* — a work that Fredrick Taylor could later say was the motivation of the scientific management movement. Metcalfe's reporting helped form a cost system that would accurately track the true cost of tooling. The record-keeping system of Metcalfe was extensive and beyond anything operating in the 1880s. In fact, it would be almost 100 years until such a system, known today as the international record-keeping system ISO 9000, would be implemented in industry. Metcalfe recorded scrap and tool failure so that problems could be solved and analyzed using data. Metcalfe's card system revolutionized machine-shop tracking, allowing for both cost control and product tracing. "Travelers" or "traveling papers" are what today's machine shops call cards. Metcalfe was years ahead of the successful "Kanban" (Japanese for card) systems used throughout the world today. The following is a summary of Metcalfe's view on record keeping from his 1885 book:

> It may be stated as a general principle that while Art seeks to produce certain effects, Science is principally concerned with investigating the causes of these. Thus, independently of the intrinsic importance of the art selected for illustration, there always seems room for a corresponding science, collecting, and classifying the records of the past so that the future operations of the art may be more effective. The administration of arsenals and other workshops is in great measure an art, and depends upon the application to a great variety of cases of certain principles, which taken together, make up what may be called the science of administration. These principles need not be formulated, nor even recognized as such, and they vary with conditions that call them forth; so that while their essence may be the same, the special rules of conduct derived from them may, in various circumstances, be widely different. Yet, for each set of conditions their character is the same, and in all they constitute what is known as our experience.[6]

The British, however, pioneered the way for Americans such as Whitney, Metcalfe, and Lee. The earliest was a Yorkshire inventor, Joseph Bramah (1749–1815). Bramah created a number of early machine tools such as the wood-planning machine and the hydraulic press. A number of Bramah's machines are preserved in London's Science Museum. Bramah was a master lock maker, and designed machines to improve production of rotating barrels and springs. One of these was a spring-winding machine, which was the forerunner of the screw-cutting machine. Manufactured locks of Bramah were world-renowned. He gained much publicity when one of his famous locks won a gold medal at the Great Exhibition of 1851. The publicity evolved from Bramah's offer of an award to anyone who could pick it. A light-fingered American achieved the feat, but it took51 hours over a 16 day period. Readers might find it easier to remember Bramah as the inventor of the beer keg pump. Another invention of Bramah was the mechanical quill cutter, which produced quill nibs that could be attached to decorated holders. Bramah's work led to the development of metal-cut nibs and the fountain pen. While Bramah's 1809 patent applied to quill cutting, he had with it invented what is called the "feed." The feed is how ink is conveyed under the nib from the reservoir to the pen point. The Bramah pen used capillary action to draw the ink. Bramah's work with the pen nib led to a series of pen nib improvements. The Bramah "engineering family" can lay claim to three generations of Bramah-trained toolmakers. Like so many Victorians, such as the Sheffield steel makers, Bramah passed his knowledge on to younger inventors. One of those students, Henry Maudslay would surpass the master.

The first example of standardized mass production by the use of machine tools goes to Henry Maudslay (1771–1831). Maudslay's first success came from his contact with the British Admiralty to produce pulley blocks for sailing ships. Pulley blocks were the key part of sailing ships for raising and lowering sails and masts. The navy actually used more than 100,000 of these blocks a year. A single man-of-war required 1,400 blocks of different sizes. As we have seen, in 1800, Marc Isambard Brunel had improved on this pulley block system, and he was searching for a machinist. Brunel teamed up with Maudslay, who was considered the best machinist of his time. The pulley block was a complex and precise product. It required machining operations such as sawing, boring, shaping, rounding, milling, and sizing. Measuring, indexing, and inspection of the product were also required. Maudslay did borrow heavily from the published work of French machinist Jesse Ramsden in 1797. Maudslay used strict accuracy and precision to ensure standardized blocks. He developed plane-surface steel gauges to check the accuracy of his manufactured pulley blocks.

Another problem of the period remained the high cost of machined products. To meet the challenge, Maudslay designed and built the first "screw-cutting lathe," or gear-driven lathe. Maybe for the non-engineer, the term 'automated lathe' makes for a better description. By means of applying different gears and cams, the lathe could be programmed to do standard and repeatable operations. At the time the simple wood screw was in high demand. The automatic screw lathe was a major breakthrough. Machine-made screws were not only cheaper but also of higher quality. Maudslay's automatic lathe and measurement system made standardized screws for the first time. James Nasmyth stated in his autobiography, "This beautiful and truly original contrivance became in the hands of the inventor, the parent of a vast progeny of perfect screws, whose descendants, whether legimate or not, are to be found in every workshop throughout the world, wherever first-class machinery is constructed." This went a long way toward the standardized assembly of furniture, boats, clocks, gunmaking machinery, and buildings. Screw production saw a boom: "British screw factories, which had annually produced less than hundred thousand gross in 1800, sixty years later produced almost 7 million gross."[7] Interestingly, Maudslay's machine was not fully automatic, but a creative American purchased one of Maudslay's machines and modified it. Cullen Whipple in 1842 developed the first fully automated screw machine, which remains today a basic screw-producing machine. Maudslay's system was in many ways 100 or more years ahead of its time. Maudslay even checked his gauges routinely to ensure that wear had not affected them. This type of calibration and gauge control has only recently become widely used with the world requirement of ISO 9000 standards. Besides the automated screw lathe, Maudslay was the first machine tool builder to use all-metal fabrication. He died in 1831 and was laid to rest in a cast-iron tomb of his own design. The epitaph on the tomb read, "eminently distinguished as an engineer for mathematical accuracy and beauty of construction." Maudslay's legacy was a new generation of machinists trained in his shops.

The Americans learned much from the Great Exhibition of 1851, returning to America with new ideas. Turret lathes and screw machines were further stimulated by the Civil War. The automatic screw machines opened up the possibility of producing precision gears and screws. These gears and screws were necessary to the development of sewing machines and typewriters. Military producers such as Remington Arms used their machining expertise to develop consumer products after the Civil War. The Vienna Exhibition of 1873 hailed the use of the electric motor to drive lathes and screw machines versus steam. Improvements in alloy steels increased the production speeds. Precision was also increased by the improvement of grinding machines. New grinding materials— silicon carbide and aluminum oxide — revolutionized the speed and accuracy of grinding operations in the 1890s. Still, it was the training of the great British machine shops that thrust the machine industry forward. Maudslay's simple shop supplied the world with a new generation of innovative machinists to apply this new technology.

One of those apprentices of Maudslay was Richard Roberts (1789–1864). Roberts had a long list of inventions that filled Machinery Hall at the Great Exhibition of 1851. The most important contribution of Maudslay was the development of other engineers, such as James Nasmyth and Richard Roberts, who worked for him. Roberts's new tools included the radial drill and the punching machine. Even more remarkable was the automation that Roberts added to these machines. Roberts adapted the Jacquard punched-card system that had automated the textile looms of Great Britain at the end of the 18th century. The punched wooden cards operated cams (gears) to "tell" the machine what type of operation to perform. This type of control was the forerunner of digitally controlled machine tools. Roberts stands as one of the most creative mechanics of the time. He improved and redesigned the mechanical differential, which was used on steam locomotives to ensure proper distribution of power to the wheels. This critical improvement allowed the locomotive's wheels to turn at the same speed and ensured smooth running on curves. Using the differential, Roberts designed and patented a smooth-running power loom in 1830. Roberts also pioneered assembly line automation in the textile industry, as noted by Cardwell: "He had, in other words, achieved batch production and was well on the way to mass production. The difference between the lines of identical textile machines in a Lancashire cotton mill of the early nineteenth century and the lines of identical automobiles in a Detroit or a Japanese car plant is more one of degree than of kind." The differential would be central to the future design of the automobile as well. Roberts, in addition, pioneered the development and use of industrial electromagnets, planing machines, locomotive design, and steam engine design.

James Nasmyth (1808–1890) was Maudslay's greatest student, and not only invented the steam hammer, but also machine tools such as the milling machine, the shaper, and drill press, as well as the engineering drawings. Nasmyth's father owned a foundry, which became an important part of his education. Nasmyth describes this training: "I delighted to watch the various processes of moulding, iron-smelting, casting, forging, pattern-making, and other smith and metal work.... I look back to the Saturday afternoons spent in the workshops of that small foundry as an important part of my education.... By the time I was fifteen I could work and turn out really respectable jobs in wood, brass, iron and steel: indeed in the working of the latter inestimable material I had at a very early age acquired considerable proficiency. As that was the pre-lucifer [matches] period, the possession of a steel and tinderbox was quite a patent of nobility among boys. So I used to forge old files into steels in my father's little workshop, and harden them and produce such first rate neat little articles in that line, that I became quite famous amongst my school companion." Later this would come in handy when Nasmyth applied for apprentice at the then famous machine shop of Maudslay. To this end Nasmyth prepared a working model of a steam engine as engineering drawings. Thanks to his expertise in tool preparation, Nasmyth became Maudslay's assistant without going through an apprenticeship.

At Maudslay's shop Nasmyth added to his expertise the art of machine making. Nasmyth also honed his skills and knowledge of heat treating and hardening steel tools. Nasmyth became a convert to Maudslay's automated or "self-acting" machine tools. Automated machines not only increased working speeds, but also reduced variation that was inherent in hand machining. Nasmyth described part of the improvement: "It gave an increased stimulus to the demand for self-acting machine tools, by which the untrustworthy efforts of hand labor might be avoided. The machines never got drunk; their hands never shook from excess; they were never absent from work; they did not strike for wages." After a few years of working on self-acting machines, Nasmyth opened his own machine shop. He started a small shop in Manchester. Maudslay allowed him to take some machine castings, which would become the basis of a series of machines in his shop. His choice of Manchester was a true blessing because of the growth of the pioneering Liverpool and Manchester Railroad. Locomotive builders such as Robert Stephenson became major customers of Nasmyth. The local economic boom gave Nasmyth a constant stream of capital to expand on. Eventually in 1836, Nasmyth began building the Bridgewater Foundry, which became a machine, foundry, and engineering complex. It was here that he built his famous steam-forging hammer that would forge the shaft of the *Great Eastern*. Probably just as important was that he used detailed drawings as a starting point for the steam hammer's design and building. These beautiful drawings were displayed with the steam hammer at the Great Exhibition of 1851, receiving much acclaim. Nasmyth also used models to help visualize possible future machines, much like Leonardo da Vinci had done. Drawing and engineering became the core of industrial art in the

An improved version of Nasmyth's original steam-forging hammer 1870 (*Scientific American*)

Victorian period. This type of graphic design would become the hallmark of engineering and machine tool design.

Maudslay's shop, like Braham's shop, served as a key training center for the education of machinists in practical heat-treating. The art of heat-treating was an ancient one practiced by sword makers in Japan and Toledo in the 14th century. The smith would heat steel white hot and hold it in the furnace to homogenize the chemistry. It was then quenched, usually in water, and tempered back to a low temperature. Different types of steel required some modifications in the heat-treating practice, which was kept secret. Apprenticing to a machine shop was the only way to gain this knowledge until the 1860s, when universities started to study heat-treating by using the microscope.

After much success with the steam-forging hammer, Nasmyth developed the steam pile driver, which revolutionized bridge building. Robert Stephenson, a railroad bridge builder, applied it first in famous bridges such as the Britannia Bridge. Nasmyth's machines became the core of industrialized England. Bridgewater Foundry was central to shipbuilding, bridge building, and the railroads. He also improved on the self-acting machines of Maudslay, improving the woodworking industries. He added to the techniques of heat-treating of tool steels. Another important invention of Nasmyth that is often overlooked was the twist drill. Drilling into metal was considered the most difficult job in the machine shop prior to the twist drill. The hole had to be drilled slowly, with the machinist stopping often to clean out the hole. The twist drill allowed the metal chips to spin out so the drilling could progress at higher speeds. Toolmakers in England formed an informal association that exchanged information. They also trained each other.

Another student of Henry Maudslay was Joseph Clement (1779–1844). Clement, like Nasmyth, was a skilled draftsman and designer. He also went to work for the famous Bramah firm and rose to chief draftsman. He then went to Maudslay's firm as chief draftsman and helped in the design of marine engines. Clement was an extremely talented designer and eventually moved out on his own. He improved and enlarged the metal planing machine. While not the first planer, it was the largest of the time. Clement's metal planing machine was a major source of income, but he was also known for his special designs. Clement crafted many special parts for building engineers, such as Isambard Kingdom Brunel. What might be considered Clement's most famous part-making effort went into a machine that never worked, but would be the beginning of the personal computer revolution of the 20th century. Charles Babbage, the father of the computer, had been working on the "difference engine." Babbage was working for the British government in the 1830s to develop a calculating machine. The machine would use a series of metal gears and revolving wheels to calculate mathematical and astronomical tables. Babbage's machine was highly specialized, and he turned to Clement to design and manufacture these needed parts. Babbage never completed the machine, but it can be seen in London's Science Museum. This type of specialized work was exactly what Clement was known for. Clement represented the engineer's engineer, customizing tooling and modeling for development work. In this sense Clement was the "craftsman" of the Industrial Revolution, building crafted parts for the engines of standardized operation. Again, Clement passed the torch on to another toolmaker.

That toolmaker was Joseph Whitworth (1803–1887), yet another apprentice of Henry Maudslay. Whitworth became the premier producer of machine tools in Europe. At the Great Exhibition of 1851, Whitworth dominated with 23 exhibits of machine tools. He added power and automation to the line of machine tools available at the time. Whitworth, however, had established himself as the toolmaker's toolmaker with his development of a machine that could measure a millionth of an inch. It was this amazing measuring machine that drew the crowds of toolmakers at the Great Exhibition. Like ISO 9000 today, Whitworth was the first to champion interchangeability of parts, standardization, gauge control, proper gauge calibration and dimensional tolerances on an international level. Here is an address of Whitworth:

> I think no estimate can be formed of our national loss from the over-multiplication of sizes. Take, for instance, the various sizes of steam engines—stationary, marine and locomotive.... In the case of locomotives and carriages I would urge the subject on the attention of our members the engineers of the great lines of railway the London and North Western, the Midland, the Great Northern, for instance. I hope they will permit me to suggest that they should consider and determine not only the fewest possible number of sizes of engines and carriages that will suffice, but also how every single piece may have strictly defined dimensions.

Whitworth's gauges were highlighted at the Paris Exhibition of 1867. There the American machinist John Sweet brought their application to American shops by studying the principles at the Exhibition. Sweet, inspired by the work of Whitworth, invented the micrometer caliper in 1873.

Whitworth's expertise in accuracy and machining led him into arms production. During the Crimean War in the early 1850s, Whitworth proposed a novel breech-loading rifle to the British government. The secret to the breech-loader is accurate, standardized screw threads. Threading a tight fit allowed for the development and advance of breech loading. British machinists were on the cutting edge of accurate screw threads, which translated into the new technology of breech loading. The British generals, however, considered it too radical. Whitworth also moved into the production of breech-loading cannons in the 1850s. Breech-loading mechanisms required machined, standardized screw threads, which were Whitworth's expertise. The Confederate army did purchase a number of 12-pounder breech-loading rifled cannons. Whitworth breech-loaders were part of Longstreet's artillery and saw action at Gettysburg. Breech loading allowed for improved speed in loading and firing. Whitworth breech-loaders could have changed the dynamics of the war if American machinists had had the capability for accurate threading.

The Braham to Whitworth machining dynasty was emulated in America with the Whitney to Colt machining legend. As the London machinists did, the early American machinists developed a type of cooperative advantage through an open exchange of techniques and tools. In turn, the Great Exhibition fostered an international exchange of information. In particular, Samuel Colt's new revolver attracted many British machinists. After the Great Exhibition, the British put together a fact-finding committee headed by Whitworth to visit American machine shops. Several others followed this committee during the 1850s. The end result was extensive and sizable orders for machine tools *from* the British, and highly improved weapon designs for the British. Historian Merritt Roe Smith noted, "Out of this complex tapestry of invention, cooperation, and diffusion can be discerned a mechanical genealogy that directly links Whitney and other early arms makers with the mass production industries of the 20th century." This type of diffusion of technology is at the root of Victorian advances. It also allowed for increased trade between participating countries such as America and England. These creative dynasties also formed the base for industrial America. Two Colt workers, Francis Pratt (1827–1902) and Amos Whitney (1832–1920), went on to form Pratt & Whitney. In 1861 Pratt & Whitney pioneered machine tools, guns, sewing machines and ultimately aircraft engines. Pratt & Whitney invented the automatic turret screw machine, which would be the muscle of the industry for the next 100 years. Out of Pratt & Whitney came the firm of Warner and Swasey! So the long tradition of evolving toolmakers continues.

The Victorian toolmakers came together with Sheffield steel makers throughout the 19th century to perfect the steel nib fountain pen. The evolution of the steel pen was not a direct jump from the quill pen. Many nib materials were being tried, such as horn, tortoise-shell, and even ruby. A pen nib needs to be flexible and elastic to write properly. Durability and wearability are the other important assets. One of the first to order steel pens for writing was Joseph Priestley. The history of the steel pen is another grail quest beyond the scope of this book. Some attribute the invention of the steel pen to an "unknown Sheffield artisan," which is fitting for any creative stream of the Victorians. Sheffield-area craftsmen advertised steel pens as early as 1803. A number of steel-pen makers seemed to be located around the Sheffield steel works. Individual craftsmen or craft shops developed these steel pens. The manufacture of the steel nib required knowledge of steel metallurgy and machining. The sequence started with the rolling of steel wire or sheet. The sheets were then made into steel blanks, which were cut with tool dies. A maker's name was then stamped on the blank and the side slits cut in. A hole was pieced in for the ink. All of this metalwork caused the small nib blank to harden. The blank had to be heat-treated by annealing in order to soften it. The nib could then be worked further by rounding it to a profile. The nib was then heat-treated again using a hardening and tempering process. Next came an acid pickle to clean the nib. Some final grounding was also required to reduce thickness in certain places on the nib. The craftsman did some final crafting, followed by varnishing to prevent corrosion and rusting. The process required a high level of steel metallurgy only known to steel makers and toolmakers. Bramah handcrafted steel pens in the 1820s, but it was a jumbled production method using his machine shop tools. The first to build a steel pen factory was Josiah Mason, a toolmaker. Mason teamed with electrochemist Elkington to gold-plate steel nibs using electroplating. In the 1850s Mason's factory employed over 1000 employees. Mason's original company was absorbed into British Pens in 1961. Mason incorporated some of the equipment developed by Bramah for quill manufacture. Another steel-pen maker was British machinist Joseph Gillott. Gillott's pens defined higher quality and particularly found popularity with artists, writers and the wealthy. His factory recorded many famous customer/visitors, such as Prince Albert, Ulysses Grant, Franz Joseph of Austria, explorer Henry Stanley, and Noah Webster.

There was a point in history, probably in the 1850s, that the amount of steel used to manufacture pens overtook the tonnage used to make swords. Steel was not the only metal nib of the Victorians. In the 1820s the platinum metal family (noble metals) was experimented with because it was highly corrosion-resistant and could outperform steel. The platinum family consisted of iridium, osmium, palladium, and rhodium. These metals were discovered at the beginning of the 19th century and, while rare, had captured the imagination of the metallic Victorians. British chemist William H. Woolaston in the mid–1820s produced a batch of rhodium-tin alloy for nibs. The batch was enough to produce about 6,500 nibs and was sold to pen makers. Another Woolaston alloy of osmium-iridium was produced but was too hard to grind. Iridium was the ideal material, but its inherent hardness limited its workability until John Isaac Hawkins invented a special high-speed lathe. The story of Hawkins and the iridium pen, however, goes back to Thomas Jefferson.

Thomas Jefferson, like Prince Albert, Napoleon III, and Abraham Lincoln, exemplified those Victorian national leaders that inspired invention and creativity. When Jefferson developed his handwriting-copying machine, the polygraph, he had worked with a little-known inventor-John Isaac Hawkins, in 1803. Jefferson loved the polygraph, but it doubled his use of quills, which was already enormous. Jefferson even experimented with a flock of geese to produce more and better quills. Hawkins hoped to improve on the polygraph to solve Jefferson's problem. Hawkins is a little-remembered inventor and mechanic, but he was prolific and profitable. In 1803 he traveled to London to study steel-pen production. Those studies actually led to the invention of one of the first mechanical pencils. The steel pen, however, did not offer the long life he was looking for. He experimented unsuccessfully with diamonds and rubies. Finally, he evaluated pens with a whole series of Woolaston alloys. Ultimately, these experiments led to the development of a powerful grinding machine to grind the extremely hard iridium. In addition, he soldered only the iridium point to a gold base nib, which cut the cost because iridium was much more costly than gold in 1834. Unfortunately, Jefferson never lived to see Hawkins's success, but the iridium-tipped pen became extremely popular in America.

One of the last Victorian pioneering toolmakers, prior to Fredrick Taylor, was Henry Robinson Towne (1844–1924), cofounder and first president of Yale & Towne Manufacturing. Towne developed not only new tooling techniques but new management techniques as well. Towne originated one of the earliest forms of gain sharing and profit sharing. Profit sharing, as we have seen, had grown in popularity in some British toolmaking shops and some early American companies. Profit sharing has it problems, which many readers can relate to. Towne realized that one department would make major gains only to be offset by other departments, thus losing the gain and profit. This could cause frustration, destroy the desire to improve, and hurt morale. Towne proposed that gain-sharing programs focus on the department or individual versus the company. Individual pay for improvement or some form of gain sharing was basic to the Victorian concept of individual motivation. It is one of those distinct differences between the Victorian Metallurgic Age and today. Tool and machine shops knew innovation was not a corporate concept, but rather the aggregate of individual efforts. The idea even goes to the heart of Victorian hero worship, which was an expression of individualism. Towne applied scientific principles to experiment with new tooling. This measuring and analysis of data allowed tooling to improve rapidly through new tools and alloys.

Fredrick Taylor, known as the "father of scientific management," first applied scientific methods to study the behavior of alloy steel before that of the machine operators. One of the limiting factors in the application of tooling is heat and wear. As tooling works, such as in planing, drilling, or cutting, heat is generated. The heat can cause the tool to soften and wear quickly. Robert Mushet had amazed the Paris Exhibition of 1865 with his tungsten tooling, which almost doubled cutting speeds. We have seen throughout the last 25 years numerous inventors improving tools with additions of chromium, tungsten, manganese, and nickel. In the late 1890s, Taylor started to set up experiments on machining feeds and speeds, using different combinations of alloys. The result was the development of "high-speed" or "red-hot-resistant" tooling, that is, tooling that became red hot with the speed of the work but retained its strength. These new steels became known as tool steels. High-speed tooling allowed cutting, screw machines, and turret lathes to run at four times the speeds of carbon tool steel. Bethlehem Steel featured these new "tool steels" at the Paris Exhibition of 1900. Fredrick Taylor and his partner, Maunsel White, won a gold medal at the Paris Exhibition. Fredrick Taylor transferred these scientific principles to the science of management. Maunsel White went on to apply the basic metallurgical science concepts of Dimitri Tschernoff and Henry Sorby to toolmaking. Sorby, of Sheffield, England, was the first to publish microscopic pictures of steel in the 1880s, and White started to look at their relationship to tool performance.

Another key to the development of tool steels was the further use of the microscope to understand the affects of heat-treating on steel structure. It was Albert Sauveur who pioneered the new field of metallography, which brought metallurgical theory and practice together. Metallography helped fuse materials science and metalworking in the field of metallurgy. Crystal structure correlated to the optimum heat-treating practices of Taylor and White, producing major improvements in the performance of tools. Sauveur today is recognized as the "dean of American metallurgists," but that title came with many setbacks. Nor was Sauveur the first. In the 1880s, French metallurgist Floris Osmond published the first microscopic studies of heat-treating. In Germany, Adolph Martens was also developing the use of metallography to study steel making. Sauveur had studied these early works at M.I.T. After graduating, he went to work with Pennsylvania Steel and took metallography from a scientific to an engineering discipline. Sauveur struggled in the early 1890s to make metallurgy part of steel making. Sauveur was a tireless promoter of metallurgy among American steel companies. Two Carnegie Steel executives and amateur chemists, Charles Schwab and William Corey, would eventually make metallurgy a department-level discipline in 1896. Other steel companies followed quickly, and Sauveur became an international celebrity. He was hired by Harvard University in 1899, and for the next 36 years brought applied science to metallurgical processing.

12

The Victorian Concept
of Management

Eli Whitney and Fredrick Taylor (1856–1915) often get the credit for the novel approaches to American management. Like the development of technology, the science of management was part of a stream of creativity. The legacy of the toolmakers goes beyond their revolutionary machines to their revolution in managing. Victorian toolmakers understood the machine-man interface as the heart of shop productivity. James Nasmyth, Joseph Whitworth, and Henry Maudslay were just three of the managers to apply a new concept of motivating employees. Nasmyth in particular pioneered these novel ideas against the popular trade unions of the 1830s. Nasmyth followed Maudslay, who in the early 1800s opposed the union apprentice system, preferring to hire "pupils." Nasmyth's philosophy was the model for generations of successful industrialists. The union apprentice approach wanted a worker to develop very specific skill sets and be paid as time in the trade accumulated. Nasmyth opposed this approach because he wanted multiskilled workers able to operate lathes, drills, boring machines and planers. He also wanted complete control over whom he might hire. Nasmyth stood by his convictions at his Bridgewater Foundry, taking a serious strike by workers later in the 1830s. Nasmyth explained it this way: "I had no difficulty in manning my machine tools by drawing my recruits from this zealous and energetic class of labourers. It is by this 'selection of the fittest' that the source of the prosperity of every large manufacturing establishment depends.... But here I came into collision with another class of workmen — those who are of the opinion that employers should select for promotion, not those who are the fittest and most [skillful] but those who have served a seven years' apprenticeship and are members of a Trades Union." Nasmyth's statement is fundamental to the argument that raged on two continents during the Metallurgic Age.

The London-Manchester machine toolmakers' fraternity was a creative center for new manufacturing approaches. Joseph Whitworth and

Henry Maudslay pioneered copartnership and profit sharing in the 1850s. Both Whitworth and Maudslay believed their success in development of tools was rooted in the development of creative employees. They resisted the in vogue training of apprentices, which was experience-based. The Whitworth/Maudslay approach focused on performance and effectiveness of training. Performance was a core virtue of the age. Bigger and better required a focus on performance. Instead of guild-controlled apprentices, they had "pupils," or students. Advancement was dependent on the student's effort, drive, and enthusiasm. The pay was actually better than in the guild apprentice system. The influx of master machine toolmakers and the explosion of tooling throughout England is testimony enough. Whitworth and Maudslay tapped into a more fundamental movement in Victorian society.

Walter Houghton described the Victorian virtue of work:

> The Victorian gospel of work, derived from both its religious and Economic life and preached the more earnestly because the idea of crisis and the idea of progress both called for dedicated action, found further support from an unexpected quarter. As the difficulties of belief increased, the essence of religion for Christians—and for agnostics the "meaning of life"—came more and more to lie in strenuous labor for the good of society. That was not only a rational alternative to fruitless speculation but also a practical means of exorcizing the mood of ennui and despair which so often accompanied the loss of faith. For these reasons, a religion of work, with or without a supernatural context, came to be, in fact, the actual faith of many Victorians: it could resolve both intellectual perplexity and psychological depression.

Even predating the early ideas of Whitworth and Maudslay was the Soho Foundry. The Soho Foundry was rooted in the Industrial Revolution of the previous century. Matthew Boulton and James Watt originally formed it. In 1800, their two sons took over the business and changed the focus from purely technical to managerial. Soho pioneered production planning, standardized components, cost accounting, employee training, incentives, production control, and employees' benefits. The Soho Foundry moved the machinist from the crafts era to the industrial era. For example, early machinists owned and maintained their own tools, where as Soho started to purchase and maintain tools for the machinist. More fundamentally, Soho tried to standardize methods moving away from the more secretive individualized methods, of the craftsman. Standard practices and methods led to a new way to motivate and pay machinists.

That customized approach was a piece-rate system of pay. The Soho system predated the more famous piece-rate systems of Babbage, Taylor, and Gilbreth. The Soho managers applied piece rate only to standardized jobs based on newly developed accounting methods. Piece rate required the worker to produce a standard number of pieces or activities to achieve a basic wage. Additional pay came for everything over the standard, motivating the employee to produce even more. While groups of men were hired at a weekly wage, Soho gave the group foreman piece-rate pay. The focus on front-line management limited the full potential of piece-rate motivation. Soho did a great deal of experimentation on pay and piece-rate systems. Management expert Claude George summarized their effort: "With many piece rates at Soho it was found that the time it took to make different size items varied more nearly in proportion to the diameter of the part than any other factor. A formula was therefore developed to express this relationship and was used for setting standards and piece rates—an example of management's use of standard data a century ahead of other firms. Soho managers, however, took great pains to make the system simple and easily understood by the workers. In all, three wages scales were used: (1) a flat piece rate for each article; (2) a piece rate varying according to size or diameter; and (3) a piece rate varying with the horsepower of the engine for fitting working gears."[8] The piece rate studies and applied systems led to another management innovation — that of cost accounting. Workers' wages could be applied to specific job activities.

The biggest advance at Soho goes against the typical characterization of Victorian industry. This was the focus on the working environment and employee morale. Worker morale was paramount to the point that management often whitewashed the foundry's walls to counter the darkness and depressing atmosphere. Workers were given subsidized housing through the company. Schools were set up for children. Christmas bonuses and family gifts were given each year. The very antithesis of Dickens's *Christmas Carol!* Boulton expressed it this way: "As the smith cannot do without his striker, so neither can a master do without his workmen. Let each perform his part well and do their duty in that state which it hath pleased God to call them, and this they will find to be the true rational ground of equality."[2] Surprisingly, we shall see that Soho was not alone in its concern for worker morale.

Another successful experiment of the early 1800s was that of New Lanark. New Lanark, Scotland, was a village of textile mills designed by Robert Owen. New Lanark revolutionized manufacture, but it was a very specialized experiment in communal manufacturing. Still, its productivity and profitability made it a point of study for Victorian industrialists. Owen summarized his concept in 1825: "Your living machines may be easily trained and directed to produce a pecuniary gain. Money spent on employees might give 50 to 100 percent return as opposed to 15 percent return on machinery. The economy of living machinery is to keep it neat and lean, treat it with kindness that its mental movements might not experience too much irritating friction."[3] Owen's paternalistic views did at times stand in stark contrast with Victorian industrialization. Still, New Lanark attracted a long array of visitors to inspect the success of Owen. Unfortunately the communal setting seemed to have masked the potential of many of Owen ideas. These noble advances of Owen were not fully ignored, but surfaced later in paternalistic corporations such as Carnegie Steel. Owen was able to co-found an American version of the experiment at New Harmony, Indiana, in 1824. From New Harmony, the Harmonists founded another industrial community at Economy, near Pittsburgh, Pennsylvania. These communal experiments again prospered, but took on unusual religious principles such as celibacy, which doomed their future growth. However, these two communities, which are restored today, left a legacy of industrial companies, including the Pennsylvania Railroad.

One of those visitors to New Lanark was Charles Babbage (1792–1871), who had undertaken a study of workshops throughout Europe. Babbage studied incentive systems, but his interest turned to division of labor. Babbage was particularly interested in the implementation of 18th century economist Adam Smith's division of labor. The style of implementation varied a great deal between shops, however, New Lanark had used division of labor in a paternalistic setting. Babbage found that to be the exception. Division of labor brought increasing gains, but it could also bring a decrease in morale. Babbage become an expert in industrial engineering principles, such as time study, to set standard production rates. Still, he did not apply the same trust levels seen at New Lanark in these studies. "If the observer stands with his watch in his hand before a person heading a pin, the workman will almost certainly increase his speed, and estimate will be too large. A much better average will result from inquiring what quantity is considered a fair's day's work. When this cannot be ascertained, the number of operations performed in a given time may frequently be counted when the workman is quite unconscious that any person is observing him. Thus the sound made by ... a loom may enable the observer to count the number of strokes per minute ... though he is outside the building."[4]

Babbage went further than Owen in the science of manufacture, and in many cases he was far ahead of his time. Babbage wanted a standardized approach to problem solving, even to the extreme of standardized forms. He may well have been the first to propose benchmarking practices when in his book, *On the Economy of Machinery and Manufactures* (1832), he called for "the use of a comparative method of studying business practices." Babbage also envisioned a centralized research-and-development function in an organization. He even proposed the use of time study years ahead of the birth of industrial engineering. One of his more interesting ideas was to study the effect of color on eye fatigue. Today we do have optic colors, such as green-yellows, to help to produce better instruments. Babbage envisioned many inventions, from the computer to automated inspection, but he seemed too mechanical in his approach to human manufacture. Clearly Babbage anticipated Taylor's work on scientific management, but lacked the cooperative approach of Owen.

Management techniques, like technology, evolved from creative streams of thought. One of the forgotten sources of management philosophy was an obscure economic lecturer at Queens College. W. S. Jevons was born in Liverpool in 1835. In 1865 he was a gold medal lecturer studying the principles of work. He published his work, *The Theory of Political Economy,* in 1871. His book predates the work of Fredrick Taylor and the Gilbreths. One of Fredrick Taylor's famous experiments, "the shovel experiment," which has been taught to management students for more than 100 years, appears in similar form in Jevons's work:

> Let us take such a simple kind of work as digging. A spade may be made of any size, and if the same number of strokes be made in the hour, the requisite exertions will vary nearly as the cube of the length of the blade. If the spade be small, the fatigue will be slight, but the work done will also be slight. A very large spade, on the other hand, will do a great quantity of work at each stroke, but the fatigue will be so great the labourer cannot long continue at his work. Accordingly, a certain medium-sized spade is adopted, which does not overtax a labourer and prevent him doing a full day's work, but enables him to accomplish as much as possible. The size of a spade should depend partly upon the tenacity and weight of the material, and partly upon the strength of the labourer. It may be observed that, in excavating stiff clay, navies use a small spade; for ordinary garden purposes a larger spade is employed; for shoveling loose sand or coals a broad capacious shovel is used, and still a larger instrument is employed for removing corn, malt, or any loose light powder.[5]

This was written a few years before Taylor started his shoveling experiments. Jevons even went beyond Taylor and the Gilbreths in proposing profit sharing and employee stock ownership in the 1870s. Like technology, management techniques were part of the Victorian creativity incubator.

Outside the machine shops, the largest early Victorian organizations were the railroads. In the 1850s, Cornelius Vanderbilt started to consolidate a number of smaller railroads in the eastern corridor. These larger organizations needed a management organization beyond the scope of the original founders' control. One-man organizations dominated Victorian industry until the growth of the railroads. The only previous large organizations had been military. Companies had been owner-driven and -managed. The railroads were unique in that they covered broad geographic areas, which made single, close supervision impossible. The organization would need a chain of command for decisionmaking and polices. Henry Poor, editor of the *American Railroad Journal* in the 1850s, called for a system that integrated organizations with information and communication.

Daniel McCallum, superintendent of the Erie Railroad, was first to apply Poor's concepts. Size, which was cherished in all aspects of Victorian endeavors, now had to be managed itself. McCallum implemented a simple organization chart to define the authority and reporting structure. To today's reader, this might seem obvious, but family-owned businesses tended to follow paternal lines. As the concept of a publicly owned, large corporation evolved, there tended to be confusion. Even today this lack of defined organization is common as family-owned organizations grow to 50 or more employees. McCallum defined operating and functional divisions as well. For example, maintenance was a specific function in the larger organization, not longer just another operation detail. McCallum also set "standard operating procedures" to make work uniform and make it easy to train new employees. He also delegated authority so problems could be resolved quickly throughout the organization. McCallum's reputation grew in the industry as an organizational genius. Certainly both Poor and McCallum are unique in that they lacked a military background, developed a grass roots approach to the new concept of the Victorian corporation. McCallum actually was hired during the Civil War by Secretary of War Edwin Stanton to run all Union railroads. McCallum demonstrated a logistical wizardry that contributed to Union campaigns, such as Sherman's March to the Sea.

Another railroad company, Baldwin Locomotive Works, applied and improved on the "armory system." Baldwin transformed the standardized manufacturing techniques into a quality-control system. The Baldwin quality-control system was far ahead of its time, implementing quality tracking, testing, and gauge control equivalent to today's international ISO 9000 process standard. In the 1860s, Baldwin's gage control program exceeded anything in manufacturing. Baldwin had not just copied the armory system but developed its own variant. Baldwin Locomotive's system of gauge and fixture control allowed it to customize locomotives while excelling in quality. In the 1880s Baldwin was shipping locomotives to Russia, South America, Central Europe and Australia, while dominating the American market. But it was quality testing that Baldwin Locomotive excelled at. In 1902 *Scientific American* did a full review of this great locomotive works. The following is *Scientific American*'s description of Baldwin's quality control:

> Before closing, a word should be said with regard to the testing department, the work of which may be said to lie at the very foundation of the excellence which characterizes the output of this establishment. All material that enters the works is subjected to both a chemical and physical test. Every delivery of plates is numbered, as is also every plate in each boiler. When a set of plates is being shipped, say from [Pittsburgh], a piece is previously cut from every

plate and expressed to the Baldwin testing department, where it is tested. The rejected test pieces are sent to the shipping clerk, and as the plate shipment comes in, the corresponding plate is returned to the makers. The boiler plate is of open hearth steel, of a tensile strength of 60,000 pounds to the square inch, and it must show an elongation of 25 per cent in 8 inches. By careful system adopted of numbering every plate in every boiler and keeping a record of the test of each batch of plates, it is possible, in case of a boiler explosion to refer to the test and obtain full data regarding the plate.[6]

This description of a quality-control system is the mark of excellence even by today's higher standards.

The military organizations and the armories of the United States were another birthplace of modern management techniques, which is not surprising because management science evolved with the need to produce more and to produce efficiently. These armories are also where the principles of science were being applied to processes. This interaction and mixing of science and men accelerated the application of science to administration. The military helped the spread of the principles of scientific management to industry. Most important was the tradition of the armories to meet and exchange information on engineering developments. This tradition had gone back to the 1830s with managers such as Rosewell Lee and Eli Whitney. In particular, the Springfield armory became a great library of new techniques. By the 1880s, engineers and managers were forming organizations to exchange information. These organizations such as clearinghouses for ideas and had been seen in the sciences in the early 1800s. The organizations allowed for debates of the new management philosophers, bringing management to the arena of science. New societies, such as the American Society of Mechanical Engineers, added conferences and open forums so papers could be presented. Books were published on the science of management by Charles Babbage, Henry Towne, Henry Metcalfe, and Fredrick Taylor. By the 1880s management science had the Victorian creative infrastructure to advance as other sciences had just as important was the interest in management science by the science fiction writers. Edward Bellamy published his look at capitalist economies in his 1888 novel, *Looking Backward*. Jules Verne also looked at manufacturing management in a number of his novels.

The Victorian concept of scientific management progressed from the early toolmakers and armories to industry in the 1880s. Fredrick Taylor (1856–1915) acted as the catalyst to apply these principles of scientific management to industrial settings. Taylor's famous theory did not just arrive; it evolved from a stream of managerial thought. Taylor had an unusual start to his career in the mid–1870s. Taylor had passed the entrance exam to Harvard's law school only to change his mind and start as a journeyman machinist. Taylor strived as a machinist for more than four years and became adroit as a skilled craftsman. The experience gave him insight into a world he had never known in coming from the wealthy class of lawyers. His working-class knowledge increased when he took a job at Midvale Steel as a pattern maker. A pattern maker crafted patterns for sand molding and ultimately casting in steel. It was at Midvale that Taylor got his first taste of supervision. It became clear to Taylor that there was a social gap between the workers and supervisors. He could be a friend in the work environment, but he was expected to be distant in social settings. This harsh contrast had deep roots in Western culture, which had maintained the Roman concept of unskilled (slave) and skilled (supervision) labor. This social standard was reinforced in Victorian society. Taylor, like typical meta-inventors, blended his insights with library research. Taylor studied and incorporated many streams of creativity in the new role of the worker. Men like Robert Owens had started to rethink this approach. A new class of management had also evolved under industrial Victorians, such as George Westinghouse and Andrew Carnegie.

Carnegie, in particular, had created hundreds of true Horatio Alger stories, and a new breed of manager. Carnegie made managers partners in his business and encouraged excellent treatment of the working class. In fact, Carnegie looked to the working class for his potential managers. Carnegie believed it was the men, not his great factories, that had made Carnegie Steel the most productive in the world. Carnegie's principles were simply an industrial application of the Golden Rule. Workers were to be treated with dignity. This was at Carnegie's core since he had risen from a poor youth in a Pittsburgh slum. Carnegie challenged his workers and supervisors to reach for lofty goals, but he rewarded them well for the effort. Social class meant nothing; performance was the only class standard. When Carnegie built his industrial Camelot at Braddock, Pennsylvania, he manned it with highly motivated, proven managers. His plant manager, Capitan Bill Jones, was a managerial genius. Jones took Carnegie's philosophical ideas and applied them to front-line supervision with the opening of the plant in 1875. At the time, British steel dominated the world market with a cheaper and higher-quality product. The British steel industry was also the technological leader. Bill Jones was to change that over the next five years. By 1879 America had matched the production of Britain, and by 1902 the United States produced 9,138,000 tons of steel, versus 1,826,000 tons for Great Britain. This success was the result of management principles rather than technology or labor costs.

The following is an example of Bill Jones's managerial beliefs in his own words: "The men should be made to feel that the company is interested in their welfare. Make the works a pleasant place for them. I have always found it best to treat men well, and I find that my men are anxious to retain my good will by working steadily and honestly, and instead of dodging are anxious to show me what a good day's work they have done. All haughty and disdainful treatment of men has a very decided and bad effect on them." Jones believed employment was a mutual contract, which was a shocking idea to the class-divided Victorians. Jones lived by his words. He was known to give money out of his pocket to help working families. When, as a manager at Braddock, his men achieved a new record, he would take them to a Pittsburgh baseball game. Jones never achieved the wealth of Carnegie but on a percentage basis out-gave Carnegie. He was always helping out his workers and their families. Probably the greatest tribute to his generosity was at his funeral in Braddock. The *Pittsburgh Gazette* reported that 10,000 tearful residents of Braddock lined the streets of his funeral procession. There are few plant managers that have ever achieved such a tribute.

Jones had fought against the industrial establishment in demanding an 8-hour day over the industry standard of 12 hours. He believed the 12-hour day to be not only brutal but counterproductive. Motivated men working 8 hours could outperform the physically and mentally worn-down men of a 12-hour shift. He had a passion for achieving goals and breaking records, which was typical of Carnegie organizations. This type of personal achievement took priority over money. While he commanded a huge salary, he gave most of it away motivating his men and helping the community. Carnegie was happy to pay Jones's demand to be paid a salary equal to the president of the United States because in five years his Edgar Thomson Works at Braddock was the most productive plant in the world. The humbled British steel industry asked Jones to come to England and tell his story.

In May 1881, Jones presented a paper to the British Iron and Steel Institute on the success of Edgar Thomson Works. These are some of the key points of that paper:

1. **"We of Edgar Thomson were compelled (being engaged in erecting the works) to listen to their wonderful stories."** Jones is referring here to the records being set at rival American steel plants of the 1870s, such as Cambria, Steelton and Joliet. Jones loved benchmarking these records for his future plant at Braddock. Edgar Thomson was built to compete with the world! World records were followed and published daily in all the departments. Bill Jones was way ahead of most of American industry; it has only been in the last 30 years that benchmarking has come into vogue. Record breaking was a passion, if not an obsession.

2. **"Esprit de corps."** Jones the sports lover modeled his managerial approach similarly. He believed in teamwork and noted it as the root to Edgar Thomson Works success. Jones further stated: "As long as the record

made by the works stands the first so long are they content to labor at a moderate rate; but let it be known that some rival establishment has beaten that record, and then there is no content until the rival's record is eclipsed."

3. **"The diversity of nationality of the workmen."** Jones believed that mixed nationalities of the workforce were part of the success. Jones described the makeup of his crews as "representatives from England, Ireland, Scotland, Wales, and all parts of Germany, Swedes, Hungarians, and a few French and Italians, with a small percent of colored workmen." He believed the multiple nationalities brought strength to an organization, but he favored the group he called "'buckwheats'— young American country boys." While Jones supported diversity, Edgar Thomson Works was segmented in "lodges," which was typical of the times. The Irish controlled the high-paying furnace jobs; the Welsh and Germans controlled the high-paying rolling mill jobs, while the Hungarians and Slavs had the low-paying labor jobs. Still, Jones protected these low-paying jobs and promoted better working conditions.

4. **"Facilities for getting the ingots out of the road."** Jones's objective was world-class throughput. Speed of output was another key to the Jones operating philosophy. Continuous flow was a hallmark of the Jones system. This emphasis led Jones to develop preventive-maintenance systems to avoid the stoppage of production.

5. **"In increasing the output of these works, I soon discovered it was entirely out of the question to expect human flesh and blood to labor incessantly for twelve hours, and therefore it was decided to put on three turns, reducing the hours of labor to eight."** Jones was the eight-hour day's biggest supporter. He was alone on this point against the industrial Victorians of the day. Carnegie allowed it at Edgar Thomson Works because of Jones but ran most of Carnegie Steel on the twelve-hour, seven-day system.

6. **"Another important matter connected with fast working is maintenance of machinery."** Here again, Jones was way ahead of his time in the concept of preventive maintenance versus breakdown maintenance.

In reality, Jones left a spirit that would be part of Edgar Thomson Works and was spread throughout industry by the likes of Fredrick Taylor. Jones's legacy is still part of America's industrial infrastructure. He developed and fostered a group of managers at Edgar Thomson Works that formed America's muscle for more than 100 years. This group was known as "the boys of Braddock," which included Charles Schwab, United States Steel's first president and the man who built Bethlehem Steel; William Cory, second president of United States Steel and a founder of Midvale Steel; William Dickson, a founder of Midvale Steel and motivational expert; Ambrose Monell, a founder with Charles Schwab of International Nickel; and many others. By 1918 this group managed almost 60 percent of America's industrial assets! The "boys," in particular, had formed the third-largest company in the United States at the time — Midvale Steel and Ordnance Company. It would be Midvale Steel that would launch the career of Fredrick Taylor, the "father of scientific management."

Midvale Steel offered a new work system so novel that some of its concepts are only now finding acceptance. This approach started with the arrival of Fredrick Taylor in the 1880s prior to the takeover of the "boys of Braddock." The Midvale Steel of the 1880s was a small ordnance manufacturer combining a machine shop with a steel foundry. It had no working system, only the old worker-class system with heavy supervision. Workers toiled for any hourly wage, having no incentive to produce more or improve their manufacture. Managerial decisions were based on intuition with no worker input. There were no standard operating procedures, which allowed each worker to develop his own methods. Supervisors often enforced the shop rules with their fists. Midvale represented the typical Victorian factory of the period. Taylor changed first before he made any changes at Midvale. That conversion Taylor attributes to joining the American Society of Mechanical Engineers and hearing about an early paper of Henry Towne — "The Engineer as an Economist." It was a short paper, but it proposed a new science that would later become known as industrial engineering.

The following embodies Towne's vision in his 1886 paper:

> Engineering has been conceded a place as one of the modern arts and has become a well-defined science, with a large and growing literature of its own, and of late years has subdivided itself into numerous and distinct divisions, one of which is that of mechanical engineering. It will probably not be disputed that the matter of shop management is of equal importance with that of engineering, as affecting the successful conduct of most, if not all, of our great industrial establishments, and that the management of works has become a matter of such great and far-reaching importance as perhaps to justify its classification also as one of the modern arts. The one is a well-defined science, with a distinct literature, with numerous journals and with many associations for the interchange of experience; the other is unorganized, is almost without literature, has no organ or medium for the interchange of experience, and is without association or organization of any kind. A vast amount of accumulated experience in the art of the workshop management already exists, but there is no record of it available to the world in general, and each old enterprise is managed more or less in its own way, receiving little benefit from the parallel experience of other similar enterprises, and imparting as little of its own to them; while each new enterprise, starting de novo and with much labor, and usually at much cost for experience, gradually develops of its managers, receiving little benefit or aid from all that may have been done previously by others in precisely the same field of work.

Towne's paper and his association with other engineers led to his basic theory of scientific management. The American Society of Mechanical Engineers, which had been formed in 1880 by Bessemer engineer Alexander Holley, served as a sounding board and showcase for Frederick Taylor's theory.

Taylor at Midvale applied the first incentive pay system. Taylor encouraged high pay through incentive programs, demonstrating that the issue is not high wages but lower unit costs. That concept was revolutionary in a society that limited pay to the working class. It was all part of Taylor's emphasis on the whole picture rather than one aspect. More important, Taylor applied scientific principles to the selection and training of employees. Training helped standardize methods and the process. He implemented the use of worker instruction cards to aid in training and standardization. Taylor pioneered cost accounting, which was critical in the evaluation of experiments and methods. Taylor's title of the father of industrial engineering stems from his use of time study to improve efficiency. He also used scientific methods to experiment with the manufacturing process. Taylor moved around the steel industry as a consultant on scientific management, working for many great companies. Ultimately, Taylor would earn the title of "father of scientific management," but full recognition came only after his death in 1915. In fact, prior to his death, the organization that had inspired his theory (Midvale) rejected it in a special committee report.

One of the other companies that Taylor revolutionized was Bethlehem Steel, which would ultimately be a proving ground for another industrial Victorian — Charles Schwab. Bill Jones of Carnegie Steel created Charles Schwab through mentoring. Schwab emerged as a product of the steel town of Braddock and Edgar Thomson Works. He was the first of a group of steel makers to be dubbed by the press as the "boys of Braddock" that would change the face of American industry. This group formed an organization known as the Carnegie Veteran Association, which met until the last died in 1938. In 1883 going to Braddock to start a steel career at Carnegie Steel was like a musician playing Carnegie Hall in New York. Jones took a poor boy and honed Schwab into a master manager. In the end, however, Schwab would go places that Jones could not have. Schwab was a polished Bill Jones, as comfortable in the boardroom as on the factory floor. Jones would not attend Carnegie's executive meetings, sending young Schwab in his place. Schwab always attributed his success to training under Bill Jones. Schwab would later recall, "I had over me an impetuous, hustling man. It was necessary for me to be up to the top notch to give satisfaction. I worked faster than I otherwise would have done, and to him I attribute the impetus that I acquired. My whole object in life then was to show him my worth and prove it." Schwab was able to take the best of Jones's and Carnegie's ideas in creating the new American industrialist. Schwab had a lovable personality that made him a very likable person. Schwab started a new era in industrial management, applying profit-sharing and gain-sharing principles.

This era was rooted in the success of the industrial Victorians such as Carnegie but broke new ground as well. Charles Schwab can be considered an industrial Edwardian, progressive on management but anchored to the past in union relations. In this view, Schwab can be considered part of a natural evolution of the beliefs of the industrial Victorians such as Carnegie, Westinghouse, Bessemer, Jones and Holley. Schwab refined the managerial gifts of Jones, as Jones had refined the gifts of Carnegie. With Schwab, for example, industrial chemistry was born. Schwab as a boy had become an amateur chemist and would later in life swap boyhood experiments with his friend Thomas Edison. Schwab applied the science of chemistry to steel making. Schwab would also be the driving force behind the introduction of production planning and industrial engineering departments. Schwab discovered an array of new uses and applications for steel. Schwab's science and persistence brought the steel I-beam into existence. He would be the founder of the two largest steel companies of the era — United States Steel and a reborn Bethlehem Steel. Most important Schwab would help maintain the long line of Braddock managers started by Jones that would transform not only the steel industry but also American industry.

Charles Schwab was the "eldest" of Bill Jones's boys of Braddock. Schwab was the architect of the formation of United States Steel. In the European press Schwab became the "Crown Prince of Steel." Schwab became a mover and shaker in American industry, being the catalyst in the formation of such companies as International Nickel. Biographer Robert Hessen said, "Perhaps the greatest symbols of Schwab's life can be seen from the center of Park Avenue in New York City, as one looks up at skyscrapers built with the Bethlehem beam." It would be some of Schwab's apprentices that would apply visionary principles to the science of management.

In the early 1900s, several of Carnegie Steel's vice presidents, William Corey, William Dickson, and Alva Dinkey, formed Midvale Steel, one of the most progressive corporations ever. The most dynamic and progressive of the three was Dickson. Dickson evolved out of Schwab's Homestead management team at Carnegie Steel. A true visionary, Dickson would take the participatory style of the "boys" to the cutting edge. As a vice president of United States Steel under Charles Schwab, he pioneered a new view of capitalism and labor. Dickson ended the barbaric use of child laborers and 12-hour work shifts. He developed a safety strategy for United States Steel that became a model for all of American industry. Dickson became a supporter of Taylor's scientific management, but it was participatory management that Dickson would be remembered for.

Dickson as a Carnegie manager had seen the power of profit sharing to motivate and inspire. He was also touched by the personal approach of Schwab with the average workers. The methodology of Carnegie and

Schwab offered motivational dynamite, but Carnegie and Schwab were limited by an anti-union paradigm. This anti-union view restricted Carnegie Steel from reaching its full potential. Dickson had seen not only the Carnegie Steel successes, but had suffered through the failures, such as the bloody Homestead strike. Dickson dreamed of merging the protection of the union with the paternal style of Carnegie's management. That dream was not restricted by management structure or unionization. He could not convince United States Steel management of his new labor ownership model. Ownership was at the heart of his new model. While the Carnegie organization had moved in this direction based on its profit sharing success, the newly formed United States Steel was under the control of conservative bankers. Dickson's opportunity would come in 1915, when he teamed up with Carnegie managers Corey and Dinkey to form Midvale Steel & Ordnance. Midvale Steel & Ordnance was a merger of Fredrick Taylor's old employer, Midvale Steel, and Remington Arms. Both of these companies had been leaders in scientific management. Dickson worked to form an employee-owned company, the first of its kind. It was to be called the "Experiment in Industrial Democracy" by labor historian Gerald Eggert.[7] It was the ultimate application of Carnegie's profit sharing, Taylor's scientific management, and Owen's industrial community. It was a Victorian vision that was years ahead of its time. Dickson's challenged the steel industry to a new structure:

> I believe earnestly that the basic idea ... offers the only solution of the proper relation of capital and labor, and wish that Judge Gary could be influenced to adopt the plan [employee ownership]. I cannot believe that the present status of labor can be maintained permanently by the United States Steel Corporation, however benevolent in intentions and practice. In effect a few men, (really three men, i.e. Judge Gary, George Baker, and J. P. Morgan) have absolute control over half of the steel industry and all persons affected by it. This is repugnant to the spirit of our institutions.[1]

Midvale Steel & Ordnance would fall by the 1920s, a victim of the economy and fearful New York bankers.

Leaders from Carnegie to Judge Gary just could not break with the idea of controlling labor. Their view, while paternal and family style, treated laborers as children. Such a paradigm embedded itself into the Metallurgic Age. It remained for a Victorian, William Dickson, to show the way around it.

13

The Copper Legend — Thomas Edison

At the Centennial Exhibition of 1876 in Philadelphia, there were several small exhibits of electrical dynamos. These revolutionary machines seemed lost in the exhibition featuring huge steam engines, yet the future was in these small generators. Two types of dynamos were on display: the Gramme dynamo of Belgium and the American Wallace dynamo. Each of the two dynamos supplied current to an arc lamp that gave a bright, bluish light. Probably the best dynamo of the time, the German Siemens, was not on display. Absent also were the developing lighting systems of Charles Brush and Thomas Edison. Still, as one explored the exhibition, the beginnings of incandescent lighting could be found in the side exhibits. One of those explorers was Edison. Edison did not invent even the incandescent lamp, which history awards to him. Moses G. Farmer, Joseph W. Swan and W. E. Sawyer all preceded Edison in inventing the incandescent lamp. The Philadelphia Exhibition hailed the creative combination of the dynamo and light bulb. Advances in lighting were reported in magazines throughout the world. In the labs of many inventors, including Edison, in 1876 were the beginnings of the final and most important engineering war of the Metallic Age. It would bring metals such as copper, tungsten, iridium and platinum to the forefront. As Edison advanced, another industrialist was preparing to enter the quest to light the world — George Westinghouse. The battle would ultimately be between these two titans of engineering. This battle would have more passion than the Kelly-Bessemer patent battle. It would also be a technological dogfight like that of Hall and Heroult over aluminum. This was to be a real war that included personalities as well technology and, more than the early Victorian streams of creativity, was a roaring river.

The inventions of lighting and dynamos to supply power were driven as a unit. The dynamo had little commercial application until the movement toward incandescent lighting. The history of the dynamo goes back to Michael Faraday. The dynamo is simply a mechanical generator of electricity. The first mechanical generator was that of Paris inventor Hippolyte Pixii. Pixii's generator was a hand-turned device rotating a magnet in fixed coils. This machine was based on the principles of Faraday and was a direct-current generator. A number of simple direct dynamos evolved throughout Europe. At the same time, Joseph Henry in America had assembled a workable dynamo. The main application of these electrical generators was in the electroplating industry. In 1876, this type of direct-current dynamo dominated the field. There were many "commercial" dynamos available and probably hundreds of lab operating models in the 1870s. Development of the dynamo had leveled by 1876; it would take a new application to spur development. The Philadelphia Exhibition had inspired the future with a dynamo powering an arc light.

Humphrey Davy had set the basis for the arc lamp when, in 1808, he connected pieces of charcoal to a battery and created a four-inch arc. Similar experiments were repeated throughout the world in the next 10 years. The arc lamp gives an extremely bright light, which limited it to special applications, such as outdoor lighting, early on. One application was in the Paris Opera. Here the arcs replaced the "limelights," which also gave extremely bright lighting. Another logical application was in lighthouses in 1858. In 1873, a Russian engineer, M. Lodyguine, used 200 arc lights to light a St. Petersburg dockyard. These Russian arc lamps were in an inert gas, so they lasted many hours. The most improved arc lamp was inspired by a visit to the Centennial Exhibition by a Russian telegraph engineer, Paul Jablochkoff. The Jablochkoff lamp used alternating current and produced a consistent, long-lasting light. The Jablochkoff lamps were used to light streets and department stores in Paris by the late 1880s. The arc lamp got a real boost from the development of an improved dynamo. The inventor of this improved dynamo was an atypical Victorian scientist in that he had a degree in electrical engineering from the University of Michigan. Charles Brush would be the first to challenge the gas-lighting systems of the 1870s and bring electrical lighting to the forefront.

Often history glosses over the invention and use of gas lighting because of its short time frame. Commercial gas lighting goes back to at least 1798, when Soho Foundry used it to illuminate buildings for night work. Soho Foundry went on to develop gas retorts for gas orders and started to canvass for orders. The Soho gas retorts, or burners, were cast iron, which was an ideal fit in their product mix. The Soho gas-lighting system was first applied to a cotton-mill operation in 1806–07. A Soho employee, Samuel Clegg, left around 1814 to improve and expand the gas-lighting industry. Clegg was able to involve local residents, such as famous art dealer Rudolph Ackermann. By 1817, London had a commercial gas-generating company —

Gas Light & Coke Company. Gas was being produced from coking coal. Gas-illuminated streets followed quickly. In 1823 more than 50 English towns had gas street lighting. In the United States, Baltimore installed gas street lighting in 1816. By the end of the 1850s, gas lighting was common in mansions and was making inroads into middle-class homes of England. The application of gas lighting changed the social structure of Victorian society. It increased literacy, reduced crime and expanded cultural events. Several historians summarized the social impact:

> The gas-mantle staved off the doom of gas as an illuminant until beyond the end of the century of which it helped to shape the social history. In the early years, Andrew Ure had claimed that gaslight adequately replaced sunlight, so that there was no moral obliquity in forcing children to work a twelve-hour day in the factories. A sounder humanitarian instinct was that which emphasized the civilizing influence of gas-lit streets, especially when they became usual in the poorer quarters of large towns. Gas-light improved the amenities of every place of public entertainment or instruction, just as it brightened the decorous domesticities of the Victorian home. Above all, it played an enormous though incalculable part in the development of the habit of reading among a population that became increasingly literate after the Elementary Education Act of 1870.[1]

The problems of gas lighting were substantial, however. Gas burned with some smoke that blackened ceilings, colored furniture, and smoked windows. This issue of dirt was resolved in the 1870s with the use of thorium and cerium oxides in the burners. There was a belief by the public that gas could cause a stupefying effect. Amazingly, in 1843, Michael Faraday was hired by the Pall Mall Club to investigate the stupefying effect of gas. Jules Verne wrote a short story, "Dr. Ox's Experiment" (1874), that played on this public fear of piped-in gas. Some of the public-relations problems were overcome in the 1870s by the introduction of other gas appliances, such as cooking stoves. A young George Westinghouse popularized gas home appliances in the 1870s, and implemented an urban distribution system for natural gas in the Pittsburgh area. The use of gas heating did not come into vogue until late in the century. The problems and slow public acceptance of gas lighting allowed an opening for the new arc-lighting systems such as those under development by Charles Brush (1849–1929) of Cleveland.

Brush in 1876 had improved on the dynamo of the period. The Brush dynamo surpassed anything on the market. Brush then turned to improving the arc lamp itself. Brush was able to automate the arc distance and maintain that length, ensuring a consistent light. He built larger dynamos and by the end of 1876, he had a 16-light machine. This unit made practical electric lighting systems commercially feasible. The Franklin Institute in 1877 tested two of Brush's competing dynamos—the Gramme dynamo and the Wallace dynamo. Brush's system had the best overall reviews. The first real application of the Brush system was a four-lamp circuit in the Philadelphia store of John Wanamaker. The success was so noted: "These lights became one of the wonders of 1878. People gathered in throngs on the sidewalks to examine them. For weeks they were talked about and a contemporary writer called them 'miniature moons on carbon points, held captive in glass globes.'" The first Brush street-lighting system was installed in Cleveland in 1879. It was a modest start with 12 arc lamps on 18-foot posts. The *Cleveland Plain Dealer* reported on April 29, 1879: "Thousands of people gathered ... and as the light shot around and through the park a shout was raised. Presently the Grays Band struck up in the pavilion, and soon afterward a section of artillery on the lakeshore began firing a salute in honor of the occasion. The light varied in intensity, when shining its brightest being so dazzling as to be painful to the eyes. In color it is of a purplish hue, not unlike moonlight, and by contrast making the gas lights in the store windows look a reddish yellow." This event was typical of the Victorian celebrations of technology, and the only thing close today are the launches of technology by Microsoft and Apple Computer. New York installed a Brush system in 1880 with similar celebrations. Jablochkoff was having similar commercial successes and celebrations in Russia. As an outdoor system, arc lighting had a strong niche, but harsh, powerful light and gases generated from burning carbon restricted its use indoors. The desire and inspiration for indoor lighting led to the development of incandescent lighting. Edison's visit to the Centennial Exhibition had convinced him to enter the development race for incandescent lighting.

Edison was far from alone in his effort to refine the incandescent electric light's filament. Decades before Edison, Joseph Swan in Britain had made significant progress. Swan was born 23 years before Edison and had begun a career as a druggist. Swan attributed his beginning studies to a second-hand edition of Faraday's *Experimental Researches in Electricity* and a lecture by Sunderland Athenaeum. Swan had also studied the patent of an unknown J. W. Starr, who patented a carbon vacuum light bulb in 1845, but died prematurely. Swan's approach emulated Edison's; in fact he performed years of trials. Still, the variations were amazingly close to each other, probably because of the publication of many experimenters. Swan, in 1878, found what he felt was a solution in the use of carbonized cellulose filaments. He appears to have beaten Edison, which is on record, with Swan giving a December 1878 demonstration to the Newcastle Chemical Society. His bulb was a vacuum tube with a carbon filament. By 1881 Swan's light bulbs were being used in Lord Kelvin's home and that of armament king Sir William Armstrong. Swan's patent in England predated Edison's as well, but Edison, with the aid of British capital, "mergered" with Swan's operation. In reality, Edison bought out the competitor and overwhelmed the propaganda machines, forcing Swan to a footnote in history.

A carbonized filament was not the only area of development, nor was Edison's work. Experimenters in Europe as early as 1840 were using platinum filaments, platinum being the highest-melting-point metal and one of the most resistant to oxidization. In 1847, Henry Draper, the world's foremost chemist-astronomer, and his father created a platinum light bulb, but it was far from commercial. Edison and Draper crossed paths in July 1877 in a western expedition by scientists to observe a solar eclipse. Edison was just ending a successful series of inventions relating to sound. It is believed that over the campfires with Draper that summer, Edison turned the full force of his efforts to incandescent lighting. Edison biographer Robert Conot noted, "Over campfires in the high plateau country, where the stars seemed to melt into the earth, conversations about nature and science lasted long into the night. The sensations of the summer were the reports from Paris of Paul Jablochkoff's 'candles,' a new form of arc light."[2] These candles would steal the show a year later at the Paris Exposition of 1878. These creative camping trips would foreshadow those of Edison, Henry Ford, and Harvey Firestone in the next century. Moses Farmer, in 1859, did perfect the platinum filament, which he used to light his farmhouse. Farmer had followed some 1850s experiments published in *Mechanics Magazine*, which also highlighted the use of another inert metal used for pen nibs—iridium. We know that Edison studied Farmer's system. Edison always started with a goal and a review of previous experiments. In Edison's own words, he outlined the approach:

> When I want to discover something, I begin by reading up everything that has been done along that line in the past — that's what all these books in the library are for. I see what has been accomplished at great labor and expense in the past. I gather the data of many thousands of experiments as a starting point, then I make thousands more.

Edison stated what had been the informal rule of all those prior. He understood that discovery comes not from a bolt of personal genius, but a review of others' genius.

Edison's path of research is worth noting. He started not with the specific electrical experiments that had been performed, but a general look at the goal of home lighting. Lighting historian Robert Silverberg summarized his start: "Since he was taking aim primarily at the gas industry, Edison collected and studied everything he could find about gas: journals of gas-engineering societies, reports of the gas companies, technical volumes. He compiled charts and tables on the economics of gas distribution. He hired an expert gas engineer to guide his investigations."[3] Edison learned some key attributes needed from his study of the gas industry. Electric lighting would need a distributed power system, not individual power sources, such as batteries at each house. It would have to be a clean system. Both gas and arc lighting were dirty energy sources. Edison set his goal precisely in his notebook (volume 184): "Object, Edison to effect exact imitation of all done by gas, so as to replace lighting by gas, by lighting by electricity." This approach of research and objective setting represents the strength of Victorian inventors. Edison, like the other Victorian meta-inventors, first tapped into the existing streams of creativity even before he set his objectives. Once he explored and evaluated the field and its potential, he set his specific objective. This type of focused research and development represented the best of Victorian creativity.

Edison embodied the model of the Victorian inventor and hero. Edison was pure Horatio Alger of the American psyche. Edison was driven and believed in himself against the opinion of his superiors. His dream would overwhelm criticism. Edison loved science from early boyhood, and he remained the passionate hobbyist. He trained himself in the principles of chemistry, electrochemistry, and electricity, but stayed away from theory and mathematical modeling. He loved libraries such as the Detroit Free Library, and built a great library to augment his Menlo Park research facility. Like most inventors in the electrical arena, Edison purchased Faraday's *Experimental Researches in Electricity* as a boy. He built a chemistry laboratory piece by piece with his earliest paychecks. As a telegrapher operator, he strived to be the fastest and best. He moved his laboratory with him to one of his first jobs on the Grand Trunk Railroad. His was the best of the Victorian amateurs, self-trained and observation-based. He hated the mathematics of his academic competitors. He was a goal setter and a visionary. He admitted studying all competitors as well as hiring specialists to help in his research. He never claimed the initial inspiration, but cherished the commercial success. He understood that history rewards commercial success more than the idea. He was part showman, which served him well in the great technological battles. Edison was a genius no doubt, but more of a driven wizard of experimentation and motivated by money, not to acquire material things but to support his creative endeavors.

Historian Robert Conot summarized the wizard of Menlo Park as "an eternal optimist who would not let himself or others consider the possibility of failure; because he was an unconventional thinker, who accumulated the resources that enabled him to transform his ideas into reality; because he charged ahead when others hung back; because he demolished the opposition and bowled over impediments. A child of the rough-and-ready universe of the Industrial Revolution, where many failed but a few succeeded spectacularly, he was the product of a unique conjunction of talent, ambition, and opportunity." [4] Edison, like many Victorian inventors, succeeded in digesting knowledge and then applying it outside current paradigms. He combined the passion of the amateur with the focus of a determined professional.

After years of research, a biographer of Edison concluded, "The Edison that I discovered was a lusty, crusty, hard-driving, opportunistic, and occasionally ruthless Midwesterner, whose Bunyanesque ambition for wealth was repeatedly subverted by his passion for invention. He was complex and contradictory, an ingenious electrician, chemist, and promoter, but a bumbling engineer and businessman. The stories of his inventions emerge out of the laboratory records as sagas of audacity, perspicacity, and luck bearing only a general resemblance to the legendary accounts of the past." This description could fit most of the great Victorian scientists, inventors and creators. Two other striking similarities can often be found in the generalized Victorian inventor: a lackluster performance and poor ratings in their early educational efforts, and strongly supportive mothers. Edison looked at a put-down by a teacher as a turning point in his life: "I found out what a good thing a mother was, she brought me back to the school and angrily told the teacher that he didn't know what he was talking about. She was the most enthusiastic champion a boy ever had that her confidence had not been misplaced."[5] Throughout his life, criticism became energy to tap into for his research efforts. Failure and struggle seemed almost inspirational to Edison. Edison took resentments and converted them into energy to achieve goals.

Edison was the ideal person to solve the remaining puzzle of the incandescent light. The technology and theory were in place; it was only a filament that was needed to close the circuit and light the world. Edison's effort is encapsulated in this review: "Thomas Alva Edison began his work on an incandescent lamp in 1877. He was by no means the first to try to develop incandescent lamps or even the first to produce one that would function. In fact, such a lamp had been made as early as 1820, and during the ensuing half-century scores of men, including, among others, Joseph Wilson Swan in England, had fashioned incandescent lamps of many designs. However, it was Edison who first developed an incandescent lamp design suitable for quantity manufacture and use."[6] Swan seems to have beaten Edison to even the carbonized filament, but Edison moved the performance into the commercial phase. Edison brought to bear not only his own brilliance but also a team of creative technicians. Edison's Menlo Park complex was a research engine to support Edison's projects. Edison ruled as a patriarch, keeping the Victorian image of the inventive hero, but in reality it was an organized research center.

Actually, it was a research colony of Edison's employees' families founded in 1876. Menlo Park, in New Jersey, had its own post office, telegraph office, railroad station, and community center. The industrial complex included a machine shop, powerhouse, and carpenter shop, as well as a library. Edison's team was always in the background, and it had technical skills and knowledge on a world-class level. Some of these were Francis Upton, Edison's mathematician; John Kruesi, a mechanic and machine-shop expert; and Ludwig Boehm, an expert glass blower and model maker. In addition, he had a number of laboratory experts. Francis Upton was not only a mathematician, but also a physicist. Edison had his own "inventor" of sorts in Charles Batchelor. Edison ran the group as his personal staff, but it freed him to focus on the critical advances. Another unique part of Menlo Park was that it attracted investors in the research effort itself. In October 1878, the research effort was incorporated into the Edison Electric Light Company. Edison competitors were now up against a research corporation with banking ties, such as those to J. P. Morgan. Still, Menlo Park was not like the departmented and hierarchical research centers that followed. Menlo Park was a Victorian research center that was more like a crafts-man's shop. Edison was the master craftsman, setting the projects and supervising the work. The employees lived and worked in the crafts com-munity. It approached the very activity of research as a craft and product in itself. Menlo Park was a creative cell, not a corporate research center. It appears strange through the prism of today's research centers. Edison, lack-ing any administrative infrastructure, ruled Menlo Park. It was autocratic and anti-creative at first sight to the modern eye, but it was a highly pro-ductive model, assuming the cell had a creative patriarch.

Menlo Park was a creative incubator mostly for Edison's ideas, and it was highly successful early on. But as the size of Menlo Park grew in the 1880s, it outgrew the patriarchal model of control. It was said that Edison "assigned projects to people, and tried to make the rounds of experimen-tal rooms like a hospital physician." Edison's success at Menlo Park worked when it was a small cell, but fell apart as the number of employees got beyond Edison's personal span of control. Edison's fault centered on his inability to delegate supervision and his lack of project management. Edi-son's model would be successfully implemented by better supervision and project review in the future. Robert Conot, Edison's biographer, describes a managerial nightmare of a growing Menlo Park: "Edison's laboratory staff had mushroomed to 120 employees—a veritable cuckoo's nest of learned men, cranks, enthusiasts, ambitious projects, he would never let go: Patrick Kenny, who had started on the autographic telegraph ten years before in Menlo Park, was still plugging away." [7] The new improvements being made by the Victorians in administration and management had not been used or understood by Edison. Edison idea of freethinking creativity unencumbered by academic bias was a plus, at least to a point. Edison's later projects, such as his iron ore separator, had huge cost overruns, drained resources, and ultimately failed. Edison's Menlo Park was ill-suited to massive trial-and-error studies.

The quest for a filament fit Edison's approach perfectly. The existing knowledge was reviewed and a clear objective defined. It would be one of endless trials and failures. In 1878, Edison was coming off his phonograph success and was in search of a new goal. He decided to work on the filament because, "Friends in science and engineering told him that the state of the art in incandescent lighting suggested that practical achievement might be near. Technical periodicals and patents also signaled activity in incandescent lighting. Such information alerted Edison to the possibility that he might solve the remaining critical problems—such as a durable filament—that would make the difference between tinkering and commercial success. He had confidence in his ability to solve electric lighting problems because, like so many professional inventors, he knew his characteristics and drew upon the experiences that had helped shape them."[8] The journey, while a fascinating story of Edison's dogged trial-and-error approach even in the face of well-defined scientific theory, might have been simplified. Was the journey necessary?

The journey would take more than 14 months and hundreds of experiments. He tried thousands of different filament materials. Edison tried many metal and oxide combinations, such as platinum, tungsten, aluminum, osmium, tantalum, thorium, iridium, and zirconium. Edison, like several before, was sure that platinum would work, but he needed a commercial success, not a scientific one. Platinum was in very short supply during the Metallurgic Age. Edison did hire a prospector to look for North American deposits of platinum. Likewise, the metals tungsten, osmium and tantalum had the properties needed, but the supply or cost ruled them out. Still, Edison continued his extensive tests on carbonized filament. He finally settled on carbonized bamboo for his filament. A lot of the credit goes to the expertise of his glass blowers and the application of a new vacuum pump. His laboratory people also developed a new type of improved dynamo. Menlo Park was a logistical and support engine for Edison's creative mind.

The search could have been streamlined by applying theory first. Nikola Tesla had always criticized Edison for his trial-and-error approach versus using theory to focus the experimentation. Edison, as brilliant as he was, failed to follow the stream of innovation that was leading to the development of the incandescent light bulb. Humphrey Davy actually was the first to attach two wires to a carbon filament (in air) to produce a glow in 1809. A German watchmaker, Henricg Globel, might hold the honor of the first incandescent light using a carbonized filament in 1854. Amazingly, Globel used what is often attributed to Edison in his worldwide search for a filament—carbonized bamboo! The use of platinum, another of Edison's favorites, goes back to the work of Warren De la Rue in 1820. The reason for the failure to commercialize his invention was the short supply of platinum, a trail that Edison would later follow. In 1875, two inventors, Henry Woodward and Matthew Evans, patented the idea of using an oxygenless bulb. It should be noted that Edison purchased the U.S. patent of Woodward and Evans in 1879. The same year both Swan and Edison put the pieces of the puzzle together.

The final success is now mixed with lore, myth and legend. Victorians liked to have a creative moment of discovery. For Menlo Park that moment would be October 21, 1879, when one of Edison's bulbs burned for 40 hours. With the "invention" released to the press and a patent submitted, Edison continued to test for better filaments. Edison applied for the patent on November 1, 1879. At the end of 1879, Edison had become the dominant inventor in the incandescent lighting field. Still, his competitors, such as those in gas and arc lighting, were digging in commercially. By the summer of 1880, Edison had discovered a very workable bamboo carbonized filament. With the bamboo filament Edison achieved a 700-hour bulb to go commercial with. The battle had only begun. Edison knew that to overtake gas, he needed an energy-distribution system. Edison's mathematician, Upton, used the mathematics of electrical theory developed in 1827 by Georg Ohm. Edison would need a high-voltage system to push current to be distributed to homes. The good news was that a high-voltage system could cut the use of copper wire by a factor of a hundred.[9] The needed dynamo prototype had been developed for his October 1879 experiments. This new dynamo could supply a constant power of 110 volts and was 90 percent efficient in converting steam into electrical energy. Most of the competition was at best 60 percent efficient in transforming rotary steam power into electricity. Edison started to build what would be the first power station for incandescent lighting at Menlo Park. The final step was to have a grand demonstration of the lighting system on New Year's Eve.

The art of the grand demonstration was Edison's forte. Edison advertised the New Year's Eve gala as the event of the century. Edison turned Menlo Park into a regional exposition. Menlo Park would be lit with more than 800 bulbs. Edison and his people would be on site to explain that the system would be cheaper to operate than gas, which had been Edison's goal. Edison opened the complex up to the public a week before to accommodate the thousands that were to come. The *New York Herald* reported on it on January 2, 1880:

> To satisfy the curiosity of the earnest inquiries on science, and to practically answer the critics and skeptics, Mr. Edison ordered the doors of his laboratories thrown wide open, that all might see and judge of his electric light ... he

set no particular night for public exhibition of the same but directed a week ago that no person who should come to Menlo Park to see the electric light be excluded from the laboratory. Availing themselves of the privilege hundreds of persons came from all quarters. During the first few days the crowds were not too large to interfere with the business of the inventor's assistants, and all went well. Every courtesy was shown and every detail of the new system of lighting explained. The crowds, however, kept increasing. The railroad company ordered extra trains to be run and carriages came streaming from near and far. Surging crowds filed into the laboratory, machine shop and private office of the scientist, and all work had to be practically suspended.

With the great demonstration, Edison ensured his place in the history books. Edison realized that this demonstration was not about pride, but financial success. Arc lighting was moving forward in Paris, Cleveland, San Francisco and New York. Swan's incandescent lighting system was gaining ground in England. And finally, the gas producers still owned the only home energy-distribution system. One of the first homes to switch from gas to electric would be New York banker and millionaire (and Edison backer) J. P. Morgan.

Edison went to extremes to make this very important home conversion successful, but it was wrought with problems. The Morgan home did make news and history as the first to be lighted entirely by incandescent bulbs. Edison installed a steam-generating plant at the rear of the home. As was the case in gas conversions, the gas tubing was used to carry the new electrical wires, so a light bulb was substituted for each burner. The power plant was extremely noisy and smoky, causing the neighbors to complain often. Edison had to use rubber padding and sandbags to muffle the noise. The power plant required a full-time "engineer." The engineer went off-duty at 11:00 p.m. often causing the lights to go out in the middle of one of Morgan's famous parties. Arrangements were made by Edison to work after-hours to ensure current. The lighting of Morgan's home represented prime publicity and goodwill for the Edison system. Off the record, Morgan was less than pleased when he said, "If it were my own, I would throw the whole damm thing into the street." Edison, however, had his victory and was ready for mass distribution.

Distribution of electrical power for Edison's incandescent lights was blocked by one of the world's oldest metals—copper. Scientists throughout the world noted that there was "not enough copper in the world" to allow for Edison's dream. The calculations of Edison's own mathematicians were depressing: "For an 8,000 lamp circuit covering nine city blocks, Edison discovered, he would need about eight hundred thousand pounds of copper at a cost of $200,000. To light all of New York City would cost billions."[10] Edison went after electrical distribution like he had approached the filament problem. In 1880, Edison took out 60 patents. Of these 32, were improvements on the incandescent lamp, 7 were on transmission systems,

Illustration from October 18, 1879, issue of *Scientific American* of Edison's first generator — "long-legged Mary Ann"

6 were on dynamos and the balance were on parts and backup equipment. The solution came with the overall system design. Instead of a tree system of lights powered by a huge dynamo, he used a main power supply and feeder subsystem. The Edison power system would be made up of a main generating station with feeder systems, similar to today's setup. The new Edison design resulted in a major reduction in the copper needed. The nine-block scenario of an 8,000-lamp circuit at a cost of $200,000 dropped to $30,000. Edison was now ready to build his first-district New York power station. Edison ordered 200 horsepower Porter-Allen steam engines and started building "jumbo" dynamos. The name "jumbo" was after the famous P.T. Barnum circus elephant. Never missing an opportunity for publicity, his first jumbo was built for the Paris Exhibition of 1881. Edison dominated Paris in 1881 as Krupp had dominated the Great Exhibition of 1851. At Paris, Edison overwhelmed his competition by having the only fully integrated incandescent lamp system. That publicity would pay off in major inroads for his system in Europe. In addition, he started laying underground wires.

Edison's fame and initial infrastructure allowed him to win the early international battles. This first effort would supply power to 900 buildings and light more than 14,000 incandescent lamps.

The first-district, or Pearl Street, system was extremely challenging to Edison. His first wires were not insulating properly and blew part of the system. He was opposed by the mayor of New York, who had ties to the gas companies. The Irish workmen had a fear of the wires that the gas companies helped perpetuate. Edison's presence was required to build confidence. Edison was driven by the need to complete the goal of a commercial system; he slept often at the work site. Equipment was often delivered late, thereby adding to the problems. Edison changed steam engine designs late in the project because of poor test performances of the Porter-Allen engines. Edison even had doubts about the system, playing down the exact date of completion to avoid any public failure. Reporters were not contacted before the system went live, although the date of September 4, 1882, did leak out. Edison watched the power come on with J. P. Morgan and a small group in

The laying of cable under the streets of New York for Edison's direct current Pearl Street station from *Scientific American*, August 26, 1882

the banker's Wall Street office. The light up went smoothly, and Edison had achieved the goal he had put in his 1877 notebook. Achieving a goal spurred Edison on to win future battles.

Even in the 1880s, the battle continued. The arc lamp manufacturers, like the Brush Company of Cleveland, continued to sell outdoor lighting systems. The public favored the softer, more natural glow of the incandescent lights for home versus the bright blue harsh light of the arcs. In Europe, through mergers, Edison gained control of the competition of Joseph Swan's incandescent lighting system. In 1883, however, Brush Electric purchased the American rights to Swan's incandescent lights. Brush Electric began installing a combined system of arc streetlights and incandescent home lights. In addition, a new incandescent competitor arose. The two professors Elihu Thomson and Edwin Houston, who had done the famous dynamo study for the Franklin Institute in 1877, formed Thomson-Houston Electric Company. Thomson-Houston started as an arc light company but with Edison's New York success moved into the incandescent market. In 1884, Thomson-Houston Electric developed a variation of the incandescent bulb. Actually, they purchased the rights of the "Sawyer-Man" bulb. It, like Edison's bulb, used a carbonized bamboo filament in a vacuum. The difference was that the Sawyer-Man bulb was first filled with gasoline vapor. Then the current was turned on, and the resulting flash would coat the filament with graphite. The flash coating extended the life of the filament. Not only was the Sawyer-Man bulb better, but it was also easier to manufacture.

In addition, Thomson-Houston Electric applied the new methods of scientific management. Edwin Rice, general superintendent at Thomson-Houston, spearheaded the application of these principles. He used a functional approach to plant organization, which was novel for the time and in stark contrast to Edison's paternal, autocratic management system. The new system allowed for focus by management specialization. Part of that specialization was the functions of cost accounting and process improvement. Cost accounting allowed for continuous cost reduction. Edison's bulb-making operation could not keep up with the manufacturing expertise of Thomson-Houston. Edison played on his name recognition to counter the competition of Thomson-Houston. In addition, Edison continued to improve on electrical transmission, introducing a three-wire system. The three-wire system actually reduces the amount of copper wire needed by 50 percent. So again Edison had found a way to far reduce the main cost of power distribution, that of copper wire. Still, Thomson-Houston grew to be the dominant company over Edison General. Historian Paul Israel outlines what the situation was in 1891: "By the end of the year it had become apparent to Edison General's mangers and investors that Thomson-Houston had a better organizational and marketing structure that allowed it to outperform its larger rival. During 1891 Thomson-Houston had earned $2.7 million on sales of $10 million and a return on investment of 26 percent, whereas Edison General earned only $1.4 million on sales of $11 million and a return of 11 percent."[11] Copper became the limiting factor to lighting the world. Throughout the later parts of the 1880s, Edison was continually running short of cash to purchase the precious copper needed for projects. To some degree Edison was selling off his ownership to get cash to buy copper. Ultimately, copper would cost Edison control of his company.

J. P. Morgan, the merger king and Edison investor, moved to end the competition by merger. Edison's weakened capital position prevented him from stopping the merger. Morgan succeeded in merging Edison General and Thomson-Houston in February 1892. The new company was incorporated as General Electric. Edison management was purged from the organization, and Edison was reduced to a board of director member. With this, the battle of the currents for Edison was over, but the war would continue. General Electric would be one of the last creations of the Victorian era — the modern corporation. The corporation was the confluence of two great Victorian creative streams — technology research and scientific management. The Edison "Invention Factory" was brought under the best of the new Victorian management concepts. In 1904, the American Institute of Electrical Engineers noted, "Never in the industrial world did organization effect a more magical change in releasing pent-up energy. Guided by master hands, electrical arts leaped into industrial preeminence; the volume of manufacture of appliances, progress of invention, public confidence in electricity, and its utilization, all took long strikes forward."[12] The wizard and his apprentices were replaced by a new type of corporate research function that was much more efficient. Ultimately, Edison's carbonized filament would yield to the superior tungsten filament in 1906.

14

War of the Currents

Edison was the master Victorian inventor, part legend and part creative dynamo. Morgan and Wall Street not the public, pushed out Edison. The public still liked to put names to scientific battles, and Edison was needed as a figurehead. The last great invention of the Victorians was to be the industrial corporation. Edison's ouster allowed for a new corporate structure to evolve. General Electric would adopt the corporate model of Edison's competitor. Still, Edison had a key role in the technology battle over AC and DC power systems. The war of the currents would now be a war of Victorian corporations. Edison would still be part of the publicity front, but the fighting would be in corporate boardrooms. Another Victorian corporate model was rising during the period led by a Victorian hero—George Westinghouse. In the end Westinghouse, like Edison, would be forced to give the fight to directors and finance men. The Victorian corporations, while a modern model for today, were unique in the way they approached technology. These Victorian corporations were aggregates of the hero characteristics that formed them. Their founders left a mark on the organization. The final war of the currents, while a battle of corporations and banks behind the scenes, was between two Victorian Titans—Edison and Westinghouse—on the front pages. Edison, Westinghouse, Ford and Firestone's ghosts are still present in the modern realities of today's companies. The Victorian creative management styles had evolved into unique organizations that were organized functionally, but retained a paternal style of manager. It is a blend that is lost in today's corporations. The Victorian corporation was creative and productive, yet it was family. There was an intense loyalty among the employees. As much as possible, there was a type of lifetime employment concept behind this loyalty. To some degree the corporation became the hero of old. The public, however, still wanted a face on these new organizations, so cartoons usually used their leaders, such as Carnegie, Rockefeller, and Westinghouse.

George Westinghouse's company best represents the classic Victorian corporation. Westinghouse was a notable Victorian inventor, having patented a rotary steam engine as a young man in 1865. Westinghouse learned metallurgy and machining in his dad's New York machine shop. Young Westinghouse in particular gained expertise in cast-iron manufacture. His first commercial invention was an improvement of the railroad "frog." The frog was a section of railway track that diverges the rails, allowing for trains to switch tracks. Westinghouse's only railroad experience was as a passenger, but, like Edison, he was a great observer. These switching frogs were the source of many train disasters, one of which Westinghouse observed as a passenger. These cast-iron frogs accepted high-impact load when trains switched, and failure would cause trains to "jump" the tracks. In addition, the frog was a high-wear item, which resulted in huge replacement and maintenance costs. Westinghouse improved the design, but more important, suggested the use of cast steel to replace cast iron. Steel has the ability to take heavy impact loading, and steel also wears many times longer. He also made the frogs reversible in that they could be turned over. This simple idea doubled the life of the frog. The final design increased the life of frogs by twentyfold. Westinghouse patented the frog and developed the new Bessemer steel plant at Troy, New York, as a supplier. The success of Westinghouse's steel frog in the late 1860s helped promote the use of Bessemer steel rails in general as a replacement for cast-iron rails. Westinghouse used Bessemer steel as a base for his foundry to produce frogs. It is yet another example of how Victorian creative streams spawned a plethora of interrelated inventions. Westinghouse, understanding the new advantages of Bessemer steel, was able to find new applications for it.

Westinghouse is credited with the first commercial use of cast steel, as biographer Francis Leupp notes: "In connection with this development, it is interesting to record the fact that Westinghouse was probably the first man, in the country at least, and possibly in the world, to produce steel castings, as the term now applies. This is an art that has ... attained a most important status in metallurgical production. He knew nothing about the subject except what he had picked up by scant opportunities for observation in early attempts to have his car-replacers and frogs made at existing steel plants; but he saw no reason why steel castings could not be produced, and so went ahead with his plans to the extent that was necessary for his particular purpose."[1] Westinghouse had of course become an expert on steel and heat treatment through much study in his dad's shop. Westinghouse also stayed close to technical publications of the time and visited the Bessemer steel works at Troy. His work on the railroad frog gained him railroad experience that ultimately made him a wealthy man.

The Westinghouse invention that revolutionized railroading and launched his Westinghouse Electric Company was the famous air brake. To fully understand its significance, let's look at a description of braking in the 1860s:

Hand braking was both difficult and dangerous. A brakeman stood between every two cars on a passenger train, and, at a point about half a mile from the next stopping place, he would begin to turn a horizontal hand wheel on one platform so as to [engage a] pair of [the car's] wheels. When he had wound the chain taut he would step across to the opposite platform and repeat the operation on the hand wheel there. No matter how skilled all the brakemen on a train might be, their work was always uneven, for no two cars would respond to the brake with the same promptness, and the slower ones would bump into the quicker adding to the hazards of the task. A freight train was harder to care for than a passenger train, because the brakemen had to ride on top of the cars in all weathers, with the liability of being knocked off by low bridge, frozen in midwinter, or, on the windy or slippery nights, missing their footing and falling between cars.[2]

Amazingly, Westinghouse actually saw a major train disaster where braking could not prevent a collision.

Westinghouse's solution to the hand-braking problem came from his interest and reading of other engineering projects. In particular, an article inspired Westinghouse in *Living Age* on the building of the Mount Cenis Tunnel. The Mount Cenis Tunnel was a seven-mile tunnel through the Italian Alps. The Mount Cenis engineers could not use steam drills directly because steam generation used air to fuel the engine, and as they penetrated miles into the mountains, air was a precious commodity. Engineers developed a steam engine compressor. The compressed air could be generated outside the tunnel and the compressed air delivered miles to operate drills. Westinghouse reasoned that compressed air could be used to distribute power throughout a train of railroad cars. Westinghouse proposed that compressed air could be generated with steam from the locomotive using a pump. Westinghouse would then supply compressed air to each car via tubes and flexible hosing. The supplied compressed air could then be used to simultaneously turn on brake shoes at each car.

Building of the air brake was a small matter for Westinghouse, but selling it became a major problem. Westinghouse had many setbacks in even getting a trial. The Pennsylvania Railroad not only rejected it, but also spread the word that it was impractical. Westinghouse teamed up with a successful Pittsburgh foundry owned by a prominent railroad manufacturer, Ralph Baggaley. Baggaley influenced his fellow railroaders and got Westinghouse in the door. Finally Westinghouse got a trial on a small Pittsburgh railroad, the Panhandle Railroad, which serviced Pittsburgh and nearby West Virginia. The trial has become legend and appears to be basically true. The trial was to take place on a run from Pittsburgh to Steubenville, Ohio. Westinghouse installed his system, trained on it, and boarded an air-brake-ready train. The *Steubenville Accommodation* on an April morning in 1869 started from the Pittsburgh Union Station. The trial was every Victorian's dream, better than any of the grand events of Edison. The train gained speed through Pittsburgh and was about to cross the Monongahela Bridge when a horse cart and driver ignored the warnings posted about the trial. The cart was stuck on the tracks as the train came out of the Grant Street Tunnel. While the train was moving at a rate of 30 miles an hour, the engineer pulled the air brake, bringing the train to a halt and saving the man's life! The trial success was a publicist's delight and an inventor's dream. In a moment, the Westinghouse air brake was famous. Westinghouse applied for a patent on April 13, 1869. Baggaley teamed up with Westinghouse to form Westinghouse Air Brake Company in July 1869. More successful trials throughout the country followed, and ultimately the U.S. Congress made the Westinghouse air brake a safety requirement. Westinghouse was a rich industrialist when Edison started his lighting experiments.

Westinghouse represented the true meta-inventor, able to move into a totally different area and create. Westinghouse's next project was a new direction as well as a learning experience. Westinghouse had built a mansion in the Homewood suburbs of Pittsburgh. Realizing the growing popularity of coal-produced gas for lighting, he dug a gas well on his property and valved it. He found a way to manage the enormous well head pressure by enlarging the pipeline. Westinghouse formed a company to exploit his new resource and drilled more wells throughout the Pittsburgh area, enlarging his distribution network. With the capital he raised, Westinghouse went into developing pipelines and distribution centers. Westinghouse in a two-year period developed and patented 28 inventions related to gas distribution. These include devices such as gas-pressure regulators, gas meters, and leak detectors. Westinghouse could go into direct competition with the coal gas producers, who were the only distributors in the early 1800s. Westinghouse was the first to distribute natural gas to industries and homes. He worked on home heating as well as gas appliances such as stoves as his distribution company grew. Ultimately, Westinghouse Electric played a key role in home appliances. His gas legacy is long forgotten, but it was this gas-distribution network that made Pittsburgh a Victorian industrial center. Westinghouse made a huge amount of money in gas distribution, but more important, like Edison's studies of the gas industry, he learned about home power distribution. Westinghouse's broad interests and universal creativity are characteristic of these Victorian inventors, such as Bessemer, Nasmyth, Edison, Whitney, Heroult, and many others. Creativity was cherished as an expertise, not specific knowledge. His brake business remained the core of his enterprises for 50 years.

In the 1880s, Westinghouse brakes had many competitors, each with a new variation or improvement. Westinghouse finally won over his competitors with electrical controls. He used electrical switches but maintained compressed air to supply the force. This brought Westinghouse into the railroad-signal industry. Westinghouse formed another company — Westinghouse Union Switch and Signal. Westinghouse developed a relationship with the Brush Electric Company through Union Switch and Signal Company to develop electromechanical controls. Westinghouse was different from Edison in that he hired consultants and engineers to work for him, working and heading up projects. Westinghouse was a natural manager, willing to delegate and to give credit. For Westinghouse it was the company that was paramount, not himself. In the early 1880s, Westinghouse teamed up with a European incandescent lamp inventor, William Stanley. In 1885, Westinghouse bought the American rights to the Gaulard-Gibbs transformer. It was but an experimental piece of equipment. Westinghouse improved it and made it practical. The transformer allowed Westinghouse to use high voltage (high pressure) to distribute electricity, and then use the transformer to set down the voltage to power incandescent lights. This revolutionary alternating current distribution would change the nature of the industry. In 1885, a Westinghouse trial generated electricity at 3,000 volts for transmission, sent the power four miles, stepped down the voltage using a transformer to 500 volts, and fed four incandescent lights. Westinghouse had used a Siemens steam engine generator to produce alternating current at high voltage. It was one of the first efforts of German engineer (and future founder of international giant electric company Siemens) Werner von Siemens. Some reporters hailed the trial as the Age of Electricity, but really it was the beginning of what would be called the war of the currents. Edison had been aware of the efficiency of high-voltage transmission but always feared the safety hazard to workers. Also, Edison was an autocratic patriarch who did not listen to his own engineers on future needs. The trial depended on the use of alternating current, which was different from the Edison direct current systems.

The war of the currents brought George Westinghouse into head-to-head competition with Edison. Edison had favored direct current because most of the dynamos of the 1870s were direct current, although Edison's light bulb could work with either AC or DC. Direct current is what today we experience in battery power. Direct current flows at low pressure or voltage, which makes it safe to handle. Edison had settled on 110 volts as safe. He seemed committed to the idea of safety. Another advantage of low voltage was that it lowered the possibility of fire related to home wires. Edison also seemed to be set in his ways, willing to accept direct current as a standard so as to work on applications. Direct current has several disadvantages. First was the limited distance that direct current could be transmitted. At 110 volts, direct current can be transmitted about two miles from the generating station. Edison designed a system of feeder stations that allowed his generating stations to serve about 16 miles. This arrangement would mean the costly building of generating stations throughout major cities. The cost of copper wire is also high in the transmission of direct current because heavy copper wire is needed for low pressure.

On the other hand, alternating current could be generated at high voltages for long-distance transmission. The disadvantages were also several. Safety was one: High voltage could kill, and this disadvantage became a battlefield in the dispute. Second, the high transmission voltage had to be stepped down, or high pressure would quickly burn out bulb filaments. The AC solution seemed necessary for a national power system, but Edison held stubbornly to direct current. One historian criticized Edison: "Ideally, the solution was to have power leave the generating station at high voltage, and then to step it down to a safe 110 volts before it reached the customer. But this was technically impossible on the Edison system. The young Edison would have looked for some way to make it possible; the middle-aged Edison simply accepted the situation as a regrettable innate flaw in his design, and left it at that."[3]

Alternating current was not new to either Westinghouse or Edison; it went back to the earliest work of Faraday. Faraday had unknowingly discovered the transformer, or "voltage changer," in 1831. The transformer was key to making alternating current usable at lower voltage. In 1838, Joseph Henry developed a step-down transformer, but with the dominance of direct current, it received little attention. In the 1880s, interest in alternating current was breaking out all over Europe. Edison's reluctance to move from direct current was not only a stumbling block for Edison General but also the nation as a whole. Edison's bias clearly slowed the creative streams in the United States on the quest for the best electrical application. In 1882, Lucien Gaulard of France and John Gibbs of England developed a step-down transformer to use between power generation and the consumer. In 1885 three Hungarian inventors improved on the Gaulard-Gibbs transformer. It was called the "Z.B.D. transformer" after the initials of the inventors. These new, more efficient transformers opened the world to alternating current. Alternating current's main advantage was in distribution and reduced copper requirements of the lines. Even Edison realized that copper might be the biggest roadblock to the growth of electrical distribution. AC did have some significant problems to overcome; first, of course, was the fact that the total U.S. electrical distribution was direct current in 1885. Second, no one had found a motor that could be run on AC. The biggest problem in the U.S. was Edison: Edison's adherence to his own original system had become something deeply personal, a matter for the emotions, not reason. He clung to his 110-volt direct current concept despite all advice to the contrary. Yet younger men all over Europe and America were excited by the new development of high-voltage AC power. High-voltage transmission suddenly was the technological darling of the moment."[4]

Edison could not stop the ultimate victory of alternating current, but he delayed it enough to maintain his prominent position in the history books. In 1885 the Brush Electric Company of Cleveland and Thomson-Houston moved into the AC field. In addition, another company, Fort Wayne Electric, started an alternating current lighting system. Westinghouse was also on the move in 1885, having purchased Gaulard-Gibbs transformers to experiment with that led to his first AC power plant in 1887. The battle lines were drawn, with big money supporting the DC Edison system at the start. J. P. Morgan and Cornelius Vanderbilt really owned Edison General, and at least initially they supported Edison's system out of self-interest. In 1886 Westinghouse fired the first public shot with the following advertisement:

> We regard it as fortunate that we have deferred entering the electrical field until the present moment. Having thus profited by the public experience of others, we enter ourselves for competition, hampered by a minimum of expense for experimental outlay.... In short, our organization is free, in large measure, of the load with which [other] electrical enterprises seem to be encumbered. The fruit of this ... we propose to share with the customer.

It was a mild beginning to what would be a dirty fight that would surpass the great scientific Victorian battle between Darwin and his opposition. The war of the currents would get personal and emotional. It would play on the public's fear and rival any political campaign. Edison had the public and Westinghouse the technology, but the final victory would be won by Morgan's moneymen. Ultimately, the best technology, the limiting cost of copper, and free competition did triumph.

Westinghouse bought up incandescent light patents, purchased transformer patents, improved on transformer technology and started building distribution networks. Westinghouse had the technological advantage with the exception of lacking an electric motor that would run on alternating current. That piece to the puzzle would come from a young Hungarian engineer, Nikola Tesla (1856–1943), a flamboyant, neurotic, and supernaturally gifted scientist that could have been a model for Mary Shelley's Dr. Frankenstein. Tesla had studied at Austrian Polytechnic, taking courses in physics and electrical engineering. He had supported his college career through gambling on cards and billiards but never graduated. He first job was with his uncle's telegraph company. He started to develop a new alternating current delivery system as early as 1881. He was hired by Edison's Paris operations. Tesla built a reputation throughout the world as an electrical wizard and troubleshooter. By 1883 he had his first working AC motor; a year later he was invited to work with Edison in America.

Tesla was brilliant and immediately contributed to Edison's machine and electrical shops; however, Edison and Tesla were exact opposites. Tesla represented the worst in an engineer — a theoretical egghead. Tesla loved to apply formulas, electrical theory and mathematics to his work. Edison hated math from the first time he read Newton's *Principia*, and he hired mathematicians to do his calculations. Personally, Tesla was well-dressed and cultured while Edison loved to play the country boy. Testa used theory in his systematic approach while Edison favored trial and error. Tesla believed alternating current was the future, whereas Edison would not even consider studying it. Tesla would replace Edison ultimately as the wizard with the new current — AC. Perhaps more important, Tesla improved on Edison's inventive approach by reducing the number of experiments using theory. Tesla recalled Edison's approach: "If Edison had a needle to find in a haystack, he would proceed at once with the diligence of the bee to examine straw after straw until he found the object of search. I was sorry to witness of such doings, knowing that a little theory and calculation would have saved him ninety percent of his labor." In April 1887 Tesla started Tesla Electric and applied for a number of AC patents. Westinghouse entered the picture by the end of the year, purchasing Tesla's patents and putting him on as a consultant.

Westinghouse was the ideal man to back Tesla. Westinghouse believed more in managing creativity than being the creative inventor. Westinghouse dreamed of alternating current dominating distribution in America, and that fit with Tesla's creative ideas. Tesla developed an entirely new distribution and operating system known as "polyphase alternating current." Tesla's formal relationship with Westinghouse lasted only a few years, but his patents would be the foundation of Westinghouse Electric. Tesla would continue the life of an eccentric but brilliant scientist. He discovered most of the devices and theories that led to radio, but Guglielmo Marconi won the initial patent battle and the commercial battle for radio by 1915. Amazingly, very much like Kelly versus Bessemer, in 1943 (the year of Tesla's death) courts reversed the decision — making Tesla the inventor of the radio! Still, it was too late to change the history books, myths, heroes, and legends, so we still honor Marconi as the inventor of radio.

By the end of the 1880s, Westinghouse had power stations, AC dynamos, and an AC motor. The price of copper was also increasing, making the Edison DC systems extremely expensive. Edison knew the problem well when he stated, "Electricity for great distances without the erection of costly conducting wires requires that the current should be of very high electromotive force, as much as 2000 to 3000 volts." The real problem and Edison's Achilles' heel was his stubbornness. Westinghouse was in a position to crush Edison's hold on the market. Tesla looked at Westinghouse as capable of the ultimate victory: "George Westinghouse was, in my opinion, the only man on this globe who could take my alternating current system under the circumstances then existing and win the battle against prejudice and money power. He was one of the world's true noblemen, of whom America may well be proud and to whom humanity owes an immense debt of gratitude." The war between Edison and Westinghouse dominated the years of 1888 and 1889.

Edison held on to a slight advantage with his DC approach based on his reputation, fear, his initial lighting systems, and lack of a good AC motor. The DC motor controlled the electric transportation system of electric trolleys. Werner von Siemens had introduced the electric streetcar or railway at the Berlin Exhibition of 1881. Siemens moved quickly to take advantage of its popularity by applying it in Berlin. Siemens sold a streetcar system to Ireland in 1884. The next five years saw the growth of streetcars in France, England and America. The electric streetcar used tracks like trains but was powered by 600-volt DC motors. The safer DC system remained popular even as most systems used AC current that was then transformed to DC for the trolley wires. The DC motor had also become popular in machine shops to drive equipment. In 1891 the engineers of Thomson-Houston developed an efficient AC motor, which would be the final nail in the DC coffin.

The Edison DC system and the Westinghouse AC system was the technological war of the Metallurgic Age. The determining factor, however, was copper. The cost of copper slowly knocked out the Edison system. The following is a summary of copper's impact:

> As the *Electrical Engineer* noted by mid-February of 1888, "If the advance in the price of copper proves to be more than temporary in its effect, one of its incidental results will be to handicap seriously the low potential system of electrical distribution [Edison's DC], in their efforts to compete commercially with the high potential system of more recent introduction [Westinghouse's AC]."

By early March of 1888, a month later, the *Journal of Engineering and Mining* was reporting,

> All the electrolytic copper in this country is now firmly in the grasp of the syndicate. There appears, in fact, nothing to prevent prices from being advanced to any figure the syndicate may wish. This unfortunate and ominous turn of events was a real blow to the Edison Electric Light Company. For instance, in the spring of 1887, the company had been putting together a bid for a Minneapolis central station powering 21,700 lights. They estimated the feeders at 254,000 pounds of copper and the main at 51,680 pounds. At seventeen cents a pound, copper costs would total $51,965. Each one-cent rise in copper pushed costs up $3,056. A three-cent rise — for copper prices were escalating steadily — would add $9,000 to the almost $52,000 price tag for copper. In painful contrast, the Westinghouse AC central plants required a third as much copper.[5]

This pressure from competitive bids would be the turning factor in the war that not even Edison could hold off. Edison held on stubbornly to his DC system, but ultimately Wall Street and the bankers would decide. Still, Edison gave it an all-out fight. This was not to be a patent fight but a down, dirty, and political fight. It was personal.

Westinghouse, moving into the New York market, was the electrical Fort Sumter. New York was Edison's turf, the site of Pearl Street Station, his first power station, and distribution network. Edison countered this time not with a technical argument, but with a fear and propaganda campaign. Edison's campaign focused on the fact that the high voltages of alternating current could kill. In early 1888, Edison brought a New York engineer, Harold Brown, to Menlo Park to study and dramatize the killing power of alternating current. Brown could also front for Edison, keeping Edison's image clean while directing the attack. Experiments started with the execution of dogs and cats. Humorous rumors emerged about the disappearance of cats and dogs in Menlo Park. Edison did legally round up stray cats and dogs throughout Orange County, and Brown staged dog and cat executions for the press. For more impact, a horse and cow were executed to demonstrate the killing power of AC. Brown even went so far as to suggest the execution of an elephant. Brown sent out "data" of accidental deaths

Generator room of Edison's Pearl Street station from *Scientific American,* August 26, 1882

throughout the country to mayors, engineers, and politicians. Edison clearly and unfairly was playing on public fear. Brown even made a call for a challenge duel between Edison and Westinghouse. Edison and Westinghouse would sit side by side as 160 volts of alternating current would pass through Westinghouse and 1,400 volts of direct current through Edison. Brown suggested victory would be determined "by who died first." Technology, however, continued to win in the marketplace.

The dirtiest part of the campaign started with Edison promoting the use of AC current in the use of state executions. Edison testified before hearings in New York to the use of alternating current. Of course Edison saw it as another way to promote the killing power and danger of AC. Westinghouse had refused to sell his AC generators for use in state executions. Edison, working through a middleman, sold AC generators to state prisons, in particular New York's Sing Sing prison and Auburn prison. The state hired Brown, to build an electric chair and set the necessary voltage. Brown made sure the press understood the importance of a Westinghouse generator in the overall system. The electric chair was ready by 1889, and a murderer became available for experimentation the same year. His name was William Kemmler. A convicted murderer, he actually volunteered, hoping to become famous in the annals of penology. Kemmler's lawyers objected, and a court battle started. Eventually Edison himself was called to testify as to the killing power of the chair. But was it humane? Some of Edison's own people suggested he back out of the project. Kemmler's lawyers argued it was "cruel and unusual punishment." Kemmler had axed an innocent woman to death, so his appeals captured no public support. The state won and prepared to go forward. Brown recommended 2,000 volts based on experiments with horses.[6] Edison believed only a 1,000 volts would be needed (or he wanted to make a point as to the killing power), and a standard 1,500-volt Westinghouse generator was purchased to be run off a steam engine in the prison's basement. The execution occurred on August 6, 1890, with about 20 scientific guests present. The official report asserted it was less painful than hanging. Unofficial reports suggested the execution had taken much longer than expected. The *New York Times* called it "an awful spectacle, far worse than hanging. This suggests that Brown's original voltage estimate should have been used, not Edison's. The term *electrocution* had not yet been coined, and the Edison people were pushing for the term *westinghousing*. This was far from Edison's greatest moment, and it is suggested that later in life Edison regretted his participation in the whole affair.

Even this public demonstration of the danger and killing power of AC current could not stop the ultimate economic victory of AC. Still, the decision was in much doubt in 1890. Edison was still the darling of the press and public. General Electric had the backing of the New York bankers such as J. P. Morgan. Westinghouse owned several large corporations, including Westinghouse Electric, Westinghouse Air Brake, and several gas companies. Westinghouse faced a number of financial and stockholder crises in the early 1890s that slowed his market penetration. He was wounded, but he was ready to fight. Westinghouse needed to expand faster because he lacked the financial backing of the Edison DC current system. He also needed a publicity victory to improve the public image of AC power systems. In 1891 Westinghouse entered a bidding war with General Electric for lighting the upcoming world's fair — the Columbian Exposition of 1893. Westinghouse had seen the power of the Edison successes in the Paris Exhibition of 1889. Edison had fought Westinghouse to a draw in the Eastern states, and Westinghouse needed to win over the cities of the Midwest to AC. Westinghouse bid the project without profit, and probably at a loss. The final bid for Edison General was $13.98 per lamp versus $5.25 per lamp for Westinghouse.[7]

Westinghouse won the bid, but he was in no position to deliver. First, he lacked the necessary bulbs, and Edison would be unlikely to sell them to him. Westinghouse was, therefore, forced to develop a new bulb. The patent battle over the use of an Edison-type bulb was tied up in court. Westinghouse quickly developed a two-piece "stopper" bulb based on patents he had purchased years before, which operated like a cork being placed in a bottle. The Edison bulb was one-piece vacuum sealed. The Westinghouse bulb would leak over time, requiring frequent replacement, driving up Westinghouse's costs even more. Westinghouse had to set up a new glass factory in Pittsburgh as well. Edison sued, charging the "stopper" bulb was an invasion of his patent and leading to more legal battles. Westinghouse had to produce more than 250,000 bulbs to ensure enough for the Chicago fair, in addition to any others he could beg or borrow. The world's fair, however, would be worth the investment.

Another problem was that Westinghouse had less than a year to build his system in Chicago. Overall, the lighting demands would be 250,000 incandescent lamps of 16 candlepower, only 90,000 of which would be used at one time; 5,000 arc lamps would also be used for park lighting. The Columbian Exposition would have about 10 times the candlepower of the Paris Exhibition of 1889. In addition, Westinghouse's dynamos would have to run an electric railway and Machinery Hall. Westinghouse applied the Tesla system, allowing for AC motors to drive electric railroads. The electric railroad was six miles long with 15 trains of four cars each. The success of the electric train helped spur the streetcar systems of many cities. The fair would operate every day for six months. The fair would log 27 million visitors, and the Westinghouse system dominated. The first benefit of the Westinghouse success was the contract of the Niagara Falls hydroelectric plant. The committee to select the design of the generation plant was headed up by the eminent scientist Lord Kelvin. The main debate centered on DC and AC currents. Westinghouse's fair success tipped the scales in favor of AC because of its ability for long-distance distribution. This would be the largest electric project ever attempted and would use Westinghouse's AC system.

In 1893 only a handful of small hydroelectric plants had been built in Europe. Actually, the first hydroelectric power plant evolved out of another technical fair. The first long-distance transmission of alternating current occurred in 1891 connecting Lauffen to the Frankfurt Electro-Technical Exposition in Germany. This was a distance of 109 miles, which was impressive. Niagara offered a massive amount of potential energy, which could be used to turn generators and produce AC current. Niagara became known as "white coal" and formed the basis of hydroelectric power generation, which even today accounts for 25 percent of electrical power generation. General Electric forced Edison out of the decision-making position and started to adapt to the new trend toward AC current. General Electric set up transmission lines to Buffalo (20 miles away) to deliver AC and DC current. Streetcars in Buffalo depended on DC current and General Electric motors. Edison was, as noted, taken out of the picture in 1892 because he had refused to be flexible as to the use of AC current. The price of copper was his Achilles' heel, and AC cost a third less in terms of copper. Furthermore, the metallic alternatives, silver and aluminum, were even more expensive. The board of directors at General Electric could not tolerate the built-in cost disadvantage of extra copper for DC generation and distribution. The headline of a New York paper summarized Edison's demise: "MR. EDISON FROZEN OUT — He Was Not Practical Enough For the Ways of Wall Street."[8] In the end, however, it was Edison's strength that became his weakness. The stubbornness and perseverance that solved the filament problem became the roadblock to the acceptance of AC. Edison just dug in more to make DC competitive against the economics and even the science of AC. This stubbornness caused Edison to commit personal assets to pay for needed copper as the moneymen tightened the noose on his company.

Edison did continue to do basic research at Menlo Park, but his time had passed. Contrary to some historians' assertions, Edison's Menlo Park was not the model for modern industrial research. Edison's Menlo Park was a pure crafts model for the research function. Edison was the master researcher with 200 apprentices. Delegation was limited, and control rested with Edison. Creativity rested in Edison's fertile mind. Tesla and other great minds had no place in such a tightly controlled environment. The Victorian corporation by the 1890s was too large for the research crafts model. The new model had been evolving in Pittsburgh. Westinghouse, a great inventor himself, was an important cultivator of creativity. Henry Prout described Westinghouse as the master of corporate creativity:

> In the manufacture of power, as in the development of transportation George Westinghouse stands amongst the apostles of democracy. He created companies and built factories in many countries. He organized, stimulated, and guided the activities of scores of thousands.... He did more, far more, for the foundation of that development than any other man who ever lived. Into it entered his imagination, his courage, and his tenacity in greater measure perhaps than into any other of his deeds.

Westinghouse's inventions were on a par with Edison's, but his ability to bring in other geniuses was his real strength. Another biographer put it this way: "Perhaps his greatest gift was his willingness to work with others. George Westinghouse was a man who liked to share ideas. He always shared credit, and he believed in rewarding talent and paying for the efforts of others. The name Westinghouse is associated with some of the greatest inventive minds of the time: Tesla, Stanley, Lamme, and many others."[9] Westinghouse was a man of deep character and always paid for the patents he used. Westinghouse didn't mind giving credit; he placed organization ahead of his own pride.

The Westinghouse concept of organization was based on innovation as a resource. Nikola Tesla stated the Westinghouse approach best: "He is one of those few men who conscientiously respect intellectual property, and who acquire their right to use inventions by fair and equitable means.... [H]ad other industrial firms and manufacturers been as just and liberal as Westinghouse, I should have had many more of my inventions in use than I now have." This was the winning attribute of Westinghouse. Here is another testimony of a Westinghouse employee who appealed to Westinghouse for ownership: "His decision was that, though the company might legally maintain its right to the inventions, he would make no move to do so, and he not only turned over to me the entire rights in the inventions, but offered me enough capital to erect and run a small factory, of which he left me in full control. I feel great satisfaction in adding that the investment proved worth while, and in bearing this witness to his fine generosity!"[10] This approach distinguishes the Westinghouse corporate view from that of Edison. Even Edison's General Electric research center adopted the Westinghouse management model for corporate research. In Westinghouse, we see the best ideas of the Victorians' scientific management come together.

Westinghouse pioneered many new management ideas. One idea he seemed to have borrowed came from a trip to England to review the industrial communities of Robert Owen — a day and half of rest per week. Westinghouse plants shut down Saturday at noon till Monday morning. He was a loving patriarch, inviting his employees to a Thanksgiving dinner at a

Pittsburgh hotel every year. Employees' picnics were common as well. West-inghouse was one of the first to offer paid vacations. He started a company pension plan and ensured profits were withheld to support this employee benefit. This pension plan went beyond anything ever seen in that it con-tinued after death for the widow and children. He ensured light work for employees hurt on the job. In 1889 Westinghouse built a model factory town called Wilmerding near Turtle Creek, Pennsylvania. It was to be a capitalist version of Robert Owen's New Lanark in Scotland. He ensured affordable housing for his employees that included indoor toilets, natural gas hook-ups, electric appliances and plumbing. Over the years he added a hospital and schools for employees. The founder of the American Federa-tion of Labor, Samuel Gompers, said of Westinghouse: "If all employers of men treated their employees with the consideration he does, the American Federation of Labor would have to go out of existence." This, of course, was the ideal of the Victorian manager: a motivated workforce, cared for, rewarded for performance, and free of unions. Westinghouse achieved what others only dreamed of. But ultimately, like Edison, he would lose his com-pany to the bankers.

15

The Automobile

The automobile, or horseless carriage, was the final project of the great Victorian engineers. It has been estimated that more than 100,000 patents combined to create the modern automobile. The history of the horseless carriage is one of definition as much as one of technology. It is a history of eliminating mammalian power for locomotion. Many historians favor the term "self-propelled vehicle." Whatever the name, such a vehicle would require the best of many streams of technology. The slowness of the automobile's development was not from its complexity, but more from lack of a vision or direction. The steam locomotive, which appeared 50 years prior to the development of the auto, represented more complexity in machinery. The story of the horseless carriage begins in 1778 with the work of French military engineer Nicholas Joseph Cugnot. To Cugnot is given the honor of the first steam-driven carriage. Cugnot was an amateur mechanic learning about the application of steam by reading. Cugnot's engine was a low-pressure steam engine. His vehicle had many limitations. It had only enough boiler capacity for 10 to 12 minutes of travel, and it generated a speed of about two miles per hour! This type of performance was certainly not a threat to anyone. One day in Paris, Cugnot lost control, and the resulting explosion of steam led it to be banned from the streets. Steam engines, because of their weight, seemed better adapted for use with rails than horseless carriages did. Jules Verne, however, did predict later in the century the use of a mechanical, steam-driven "elephant" in his novel *The Steam House*. A number of inventors did develop "road locomotives" similar to Verne's fiction, but weight, water requirements, and fuel requirements made them impractical for mass use. The consensus of science fiction writers, dreamers and inventors was that the horseless carriage would be steam-powered. At the turn of the 19th century, steam-powered cars did outnumber gasoline-powered cars due to the strength of the steam paradigm. The idea of personal freedom in transportation also lacked a vision for most of the century.

Automotive historians point to the invention of the bicycle in the mid–1800s as a necessary psychological breakthrough. In particular, the bicycle opened up the idea of flexibility and personal freedom. In the 1980s automotive historian John Rae noted:

> It has been the habit to give the gasoline engine all the credit for bringing in the automobile — in my opinion this is the wrong explanation. We have had the steam engine for over a century. We could have built steam vehicles in 1880, or indeed in 1870 [if not much earlier]. But we waited until 1895.
>
> The reason why we did not build road vehicles before this, in my opinion, was because the bicycle had not yet come in numbers and had not directed men's minds to the possibilities of long distance travel over the ordinary highway. We thought the railroad was good enough. The bicycle created a new demand which went beyond the ability of the railroad to supply. Then it came about that the bicycle could not satisfy the demand it had created. A mechanically propelled vehicle was wanted instead of foot propelled one, and we know now that the automobile was the answer.[1]

Aside from Jules Verne, there was also a lack in the science fiction visionaries in predicting the development of the horseless carriage. The technological and creative stream was in place but had no direction for the evolution of a product. This was different from so many of the Victorian inventions, where the applications were awaiting the process or invention. Zeppelin, for example, had designed his aluminum-framed airship 20 years before there was enough commercial aluminum to build it.

The bicycle, while primarily a psychological breakthrough, did contribute some key mechanical devices for the automobile such as Richard Roberts's differential, ball bearings, and chain drives. The role of the bicycle in the mechanics of the automobile is summarized in the book *Engineering in History*: "While the bicycle does not present complicated engineering problems, the history of its development shows an important and curious relationship to the early history of the automobile. The first mass producer of the safety bicycle, James Starley, reinvented the differential gear now so necessary for the automobile. The chain drive used in early automobiles was first used on bicycles. The bicycle thus played an important role in the invention and evolution of the automobile." The "safety" bicycle refers to a model that used a sprocket and chain drive, allowing the driver to brake. The bicycle is also credited with the improvement of roads, at least in urban areas. This improvement in roads was necessary for automobiles, as early pioneers found out. Bicycles and road improvement fostered an organization known as the League of American Wheelmen (in Europe, the Cyclist's Touring Club). A forerunner of AAA, it became a political force for road improvement in the 1890s.

If we eliminate steam locomotion as a main branch of the automobile's development, we can start with Samuel Morey (1762–1843). Morey was a generalist and engineer of the Victorian era, patenting inventions in navigation, agriculture, and steam generation. Morey's main contribution was the development of the internal combustion engine. This probably was not the first internal combustion engine, since evidence of earlier British patents can be found. Some of these go back to the 18th century, but the railroad paradigm masked any potential use. Again, history notes the lack of use stalling its development: "With the internal combustion engine, as with so many inventions, the fundamental idea was conceived long before there existed either the means of putting it satisfactorily into practice or a strong incentive to develop them."[2] Again, even the science fiction writers could not supply a use to focus this stream of creativity. Morey's patent was filed in 1826 with the following description: "This discovery consists of a machine or engine, for mechanically mixing or preparing gases or vapors with atmospheric air, in such a manner as to render them highly explosive, and exploding the same, thereby forming a vacuum, by means of which a power is derived applicable to nearly every mechanical purpose."[3] Morey vaporized a liquid fuel mixture of turpentine, alcohol, and/or camphor. His engine was made up of two cylinder-piston combinations. His first engine produced rotary power, turning a flywheel. He demonstrated his engine in Baltimore and Philadelphia in the late 1820s. Maybe Morey's biggest contribution was to conceive of the automobile. In Philadelphia, Morey almost made his first and last self-propelled demonstration trip using the engine. Morey started the carriage on Market Street but lost his grip, and the vehicle left on its own. It eventually crashed into a brickyard. Automotive engineer Horst Hardenburg concluded, "It was the first witnessed — even if involuntarily unmanned — automobile run in America."[4]

The first actual self-propelled successes had to wait until the 1860s in Europe. One of these was from the French inventor Etienne Lenoir, and the other an Austrian inventor, Siegfried Marcus. Marcus certainly has a claim to the gasoline motorcar. Reportedly, Marcus put a four-stroke engine on a handcart in 1864. An 1874 version of this car is preserved in Technical Museum in Vienna. The amount of activity occurring in motorcars from 1860 on was amazing. Most of the designs were variations of the bicycle and tricycle. The paradigm of the bicycle as the future of self-propelled vehicles was deeply imbedded. The invention of the motorcycle would actually precede that of the automobile. The tricycle shaped early car designs as well.

Morey had, however, started the concept of the modern automobile, and that gave the direction to the river of creativity building. In the 1870s, various experimental vehicles were appearing. One was a direct result of the Philadelphia Centennial Exhibition of 1876. American inventor George Brayton exhibited an oil-based internal combustion engine that he had patented in 1873. A Pittsburgh bus operator, James Faucett, contracted Brayton to build a street bus. The Pittsburgh vehicle operated successfully but lacked power to climb the hilly terrain with any speed. The Brayton engine used on the Pittsburgh bus was the first to use a distillate of oil known as "gasoline." The Brayton engine took the leadership position throughout the world. In 1878, submarine inventor, John Holland, used a Brayton engine in his first submarine. More important, Brayton licensed his engine to German engine producers Gottlieb Daimler, Nikolaus Otto and Eugen Langen. The Brayton engine thus became the basis for the emergence of a European self-propelled car. The most interesting part of Brayton's success at the Centennial Exhibition was that of attorney George Baldwin Selden. Selden made a great deal of money in auto making without ever having built a car. Selden, on seeing the Brayton engine at the exhibition, drew up a patent applying the engine to a horseless vehicle. He never built the vehicle but sued for patent infringement throughout the world, successfully collecting royalties. So many auto producers paid off Selden that in 1895, he was granted a patent for "his invention."

The 1880s saw an explosion of horseless carriage inventions. If you drive through Plymouth, Pennsylvania, you will pass a historical marker claiming the home of one of America's first automobiles. Plymouth was the home of machinist Sephaniah Reese, who in the late 1880s built a three-wheel car. The car used a tiller like that of Henry Ford's first car to steer the vehicle. Reese advertised for years that he would custom-build cars, but it appears only one was ever produced. The Reese car was a gasoline-powered vehicle. Reese appears to have beaten the competition to the first gas-powered internal-engine car, but he was not the first to patent it. The honor of first "reliable" car produced goes to gasoline engine producer Karl Friedrich Benz of Mannheim, Germany. Benz did receive the first patent for a gas-powered internal-combustion car. It is again the story of Bessemer and Kelly that is the paradigm of Victorian invention and the development of technology. Benz's car was a three-wheeler, but it pioneered many modern automotive components. The Benz car of 1885 had the mechanical operations of today's automobile. Benz was the first to apply the differential gear to the wheels, allowing for equal power distribution. Benz's car also had electric ignition, which was an original invention of Benz. Benz also designed the first water-cooled radiator system. About the same year, Gottlieb Daimler put a gasoline engine on a bicycle, pioneering the first motorcycle. In 1888, Daimler built his first automobile. Daimler's car had several improvements over Benz's. Daimler had a four-speed drive versus one for Benz. He added a reverse gear and a gearbox to allow for speed changes. The Daimler car also was first to use Dunlop pneumatic tires. Benz borrowed from Daimler, producing a durable and reliable model, which he featured at the Columbian Exposition in Chicago in 1893. The Benz car would inspire American mechanics like Frank Duryea and Henry Ford.

The internal-combustion automobile is a complex invention with a complex history. Reese drove one first, Benz was first to patent one, Duryea won the engineering title as first, and Ford holds popular history's title. Frank Duryea (1869–1967) grew up on an Illinois farm, where he learned mechanics. As a young man he worked with his brother, Charles Duryea, at the Owens Bicycle shop in Washington DC. In 1888, the Duryea brothers moved on to the Ames Manufacturing Company in Springfield, Massachusetts. Ames was one of America's premier factories and toolmakers. Inspired by an 1891 *Scientific American* article on the Benz motorcar, the brothers started a major research effort. The brothers spent months researching gasoline cars and engines at the Springfield library. In a small shop, the brothers started to experiment with differential gears on tricycles. Engines, such as the Benz and Atkinson four-cycle gas engines, were also experimented with. These engines lacked the refinements of today, but they are basically the same. The experiments developed into their own "horseless carriage project." In 1891, putting the cart before the horse, they purchased their first component — a $70, heavy, second-hand buggy. Frank Duryea took on the whole project, renting space at Springfield's Russell & Sons machine shop.

Duryea's project was a full-scale crafted motorcar. In this respect, Duryea broke new ground. He worked laboriously, hand-crafting parts and fitting each component. Duryea forged his own axles and transmission shafts. Gears and wheel bearings were machined. Using his four-wheel buggy as a base, Duryea added a suspension and springs. He designed a friction transmission, breaking from the chain-gear paradigm of the bicycle. He contracted Ames Manufacturing to produce a cast-iron cylinder for the engine. The engine was based on the four-stroke German engine but was a novel two-piston setup. Ignition problems plagued the engine operation for months, requiring a redesign. In 1893, engine failures continued to hold up Duryea's development of a practical motor. The first successful but limited drive occurred on September 21, 1893. This first Duryea car is today in the Smithsonian. In the middle of these problems, Duryea took time to go the Columbian Exposition in Chicago. He spent days taking rides in the Daimler motorcars on exhibit as well as rides on the new Sturges motorcycle. Duryea returned to Springfield motivated to make his vehicle a commercial success. He had several major hurdles to address. The first was that he needed a muffler because the noise was "terrific beyond words." Second, he needed a carburetor to smooth the starting and running of his engines. The first commercial version of the Duryea car was ready in late 1894, but there was no market and almost no interest.

Duryea would, however, seize a new Victorian-style opportunity. In the fall of 1895, the *Chicago Times Herald* announced a Thanksgiving Day

"Motocycle Race." As part of the promotion, a prize of $500 was offered for the best name for the horseless carriage. The public hated the term "automobile," favoring "motocycle," which was awarded the prize. The race was an international event with a prize of $5,000. There were more than 75 entries, including several Benz cars. The day of the race there were only six vehicles, two of which were electrics. Of the four gasoline entries, three were Benz cars, and the other was a Duryea. It was a snowy day favoring the better traction of the Duryea car. Duryea triumphed that day. The official report read, "Three and one-half gallons of gasoline, and nineteen gallons of water were consumed. No power outside the vehicle was used. I estimate that enough power was used to run the motor 120 miles over smooth roads. Our correct time was seven hours and fifty-three minutes. We covered a distance of 54.36 miles—averaging a little more than seven miles per hour."[5] Interestingly, one of the electric entries, the Electrobat, was given a gold medal with the following citation: "ease of control, absence of noise and vibration, cleanliness, and general excellence of design." The month of the race the first edition of the industry paper *The Horseless Carriage* was published with the following prophetic words:

> Those who have taken the pains to search below the surface for great tendencies of the age, know that a giant industry is struggling into being. It is often said that a civilization may be measured by its facilities of locomotion. If this is true, as seems abundantly proved by present facts and the testimony of history, the new civilization that is rolling in with the horseless carriage will be higher civilization than the one that you enjoy.

Certainly, the Duryea had set the stage for Henry Ford, but actual production of horseless carriages suggested a different story. "In 1900, for example, the automobile industry produced 4,192 cars; of these, 1,681 were steam powered, 1,575 were electric, and 936 were gas powered."[6]

Epilogue

The Victorian period represents an explosive age of invention. The period has no equal except possibly 1960 to 1980, in which the 20th-century streams came together to conquer the moon and create the PC revolution. This period has many similarities to that of the Victorian Metallurgic Age. First there was a focusing of creative genius through science fiction, world's fairs, the press, and hobbyists. Science fiction reached a peak in 1960 along with a revival of Jules Verne in the movies. Space exploration was the major focus of the science fiction of the period, leading the way to a national goal in the 1960s. The space race followed a lot of Verne's original predictions and design. This was not surprising since most of the American space scientists were products of the 1950s Verne revival. Besides the clear goal development in the Victorian period, there was a democratic and catholic approach to information streams. This allowed amateurs and hobbyist to play an important role along with scientists. Hobbyists helped make the technical fairs successful. Hobbyists were promoting rocket modeling in the 1960s as well. The PC revolution of the 1980s followed a similar path. The leading pavilion at the 1964 New York World's Fair was IBM, which featured the giant IBM 360 computer. Again, hobbyists led the way in computer development with kits. Hobbyist science groups such as the "Homebrew Computer Club" inspired many inventors, including Steve Jobs and Bill Gates. The Metallurgic Age also had many "homebrew" clubs dedicated to chemistry, metallurgy, geology, and the intermixing of hobbyists and scientists was the norm. The scientific associations supported this mix because of the synergy and creativity that evolved from it. Even today, where mixing occurs, such as in astronomy, computers and robotics, we see major advances in innovation.

Currently, astronomy follows the Victorian model best with a blend and cooperation among amateurs and professionals. Comet hunting, in particular, represents the advances of such a cooperative effort. Amateur David Levy teamed with professional astronomers Gene and Carolyn Shoemaker to predict and observe comets colliding with Jupiter. It is in astronomy that we see the passion of the amateurs fused with the scientific knowledge of the professional. Levy described the blending: "While amateurs are not doing astronomy for a living, it's certainly not just a hobby for us. It is part of our nature.... If you are a professional astronomer, you are doing astronomy as a daily activity to earn money. There's nothing wrong with this, and the fact you're a professional astronomer doesn't stop you from being an amateur, too." This describes also the nature of Victorian science and engineering. In the Metallurgic Age, there existed a mix of amateur and professional metallurgists moving together. Pneumatic steel making required both the professional chemists, such as Kelly, Mushet, and Gilchrist, and the amateur inventors, such as Nasmyth, Bessemer, and Jones. It is a model repeated with the development of aluminum, the telephone, and the electric light bulb as well.

Science fiction, exhibitions, fairs, associations, and the press focused on the advance of engineering in a way that has been lost. The exhibitions, in particular, motivated the masses. The exhibitions stirred national pride and competition for industrial advancement. The great meta-inventors of the age were constant visitors and exhibitors. Many, such as Nasmyth and Bessemer, were daily attendees. Others, such as Morse, Edison, and Bell, launched their commercial marketing campaigns at the great fairs. Krupp's great exhibits inspired decades of steel and ordnance development. The steam hammer, telegraph, telephone, incandescent light, electric motor, automobile, locomotive, machine tools, and electric railways, to name but a few, were all introduced at exhibitions. The exhibitions fired the imaginations of the press and science fiction writers such as Jules Verne, who in turn fired the imaginations of the inventors. National leaders such as Napoleon III and Prince Albert, influenced by the technology of the exhibitions, encouraged inventors with awards, medals and financial rewards. Even critics of the social evils of the era were awed by the great technology of the exhibitions. The Victorian exhibitions continued long into the 20th century, with the last great one being the 1964 New York World's Fair. The exhibitions were technology for the masses.

The focus of Victorian science fiction was the education of the masses. Jules Verne, of course, stands out in his ability to blend fiction and science. But just as important were the wonder, awe, and inspiration that science fiction propagated. Science fiction many times set the direction for the powerful streams of creativity. Verne, in particular, not only created dreams but also set goals for science. Many of Verne's predictions can be considered self-fulfilling prophecies because his novels framed paradigms for engineers and scientists. Of course, the other aspect of Victorian science fiction was inspirational. Victorian science fiction was clearly scientific romance. Fantasy was never in vogue because it was outside the realm of possibility.

Verne and writers like Wells and Conan Doyle created heroic models for scientists and engineers. Even the mainstream writers contributed: Mark Twain, Jack London, Rudyard Kipling, and Edgar Allan Poe. Most, however, never achieved the blend of science and romance that Verne did. Verne's romances have achieved a timeless character far beyond the dated technology. H. G. Wells never reached the level of popularity of Verne because often Wells lacked the hard science to support the dream. Verne's trip to the moon was a reference for more than 100 years for scientists, hobbyists, engineers, and dreamers. It foretold the use of aluminum, the existence of weightlessness and the use of retro rockets. Verne and the others also supplied a new group of heroes—scientists, professors, and engineers.

The creative inventors of the Victorian Metallurgic Age were not technical specialists, but generalists. Most were mechanically oriented, but some such as Thoreau were primarily artists. Engineering historian Henry Petroski noted, "The story of Thoreau is instructive because it is a reminder that innovative and creative engineering was done by those who were interested in a wide variety of subjects beyond the technical. Whether or not they had college degrees, influential early-nineteenth-century engineers could be a literate lot, mixing freely with the most prominent contemporary writers, artists, scientists, and politicians. And this interaction hardened rather than softened the ability of the engineers to solve tough engineering problems."[2] It went both ways. Paxton, the designer of the Crystal Palace of the Great Exhibition of 1851, was an accomplished and published gardening expert. The great meta-inventors, such as Westinghouse, Bessemer, Edison, and Nasmyth, moved freely between diverse fields of knowledge to invent and create. Westinghouse, for example, had inventions in the gas-distribution industry, railroad industry, appliance manufacturing, incandescent lighting, and electrical power distribution. The inventor of the steam hammer, James Nasmyth, was an amateur astronomer and accomplished artist. Samuel Morse, inventor of the telegraph, was a famous and accomplished painter. Alfred Nobel was also an artist. Many writers of the time, such as Henry Thoreau and Mark Twain, were also inventors. This mixture of artist and engineer seems characteristic of the meta-inventors of the age. Michelangelo and Da Vinci appear to be the earlier models for Victorian inventors. The linkage is related to the very nature of creativity.

Amateurism and diversified interests helped to produce creativity in the Victorian era. The diversified backgrounds not only helped break paradigms, but also added a balance and a spirit to inventions. This creative mix led directly to the idea of "industrial arts." The blend promoted a popularity and interest in industry and inventions. Steve Jobs more recently talked of the same blend of art and invention in the development of Apple's

Macintosh. Jobs was proud of his Macintosh team, which was made up of artists, biologists and scientists who happened to be outstanding computer programmers. It's what gave us the beautiful fonts of the word processing programs. This represented the routine nature of creative streams in the Victorian era. Art and engineering were the beneficiaries of the Victorian burst of creativity. Diversity became a cherished virtue to the Victorians. Creativity became a craft in itself that would be a tool in any field. The meta-inventors of the Victorian era, such as Bessemer, pioneered inventions as diverse as sugar presses, gold paint, steel making and stamp-cancellation apparatuses. Their skill consisted of the ability to tap into a stream of knowledge and creatively apply it. Even non-inventors, such as Lincoln, Napoleon III and Prince Albert, directed national resources to create, and directing became an important part of the craft of creativity.

The touchstone of the Victorian meta-inventor, leader, and manager was the ability to synthesize by bringing creative streams of knowledge together. Professor Richard Florida summarizes it: "Einstein captured it nicely when he called his own work 'combinatory play.' It is a matter of sifting through data, perceptions and materials to come up with combinations that are new and useful. Creative synthesis is useful in such varied ways as producing a practical device, or a theory or insight that can be applied to solve a problem, or a work of art that can be appreciated."[3] Florida supports the view of Murray, who links creativity not to intelligence but cognitive ability. The Victorian culture favored cognitive ability over intelligence. Economic rewards favor cognitive skills over intelligence. This viewpoint might also explain the close linkage between art and engineering in the Victorian Metallurgic Age. Engineers also are favored by cognitive ability over scientists who depend more on raw intelligence, but even meta-scientists such as Einstein favored cognitive skills. These cognitive skills allow for breakthroughs by thinking "outside the box." This emphasis on the importance of cognitive skills is demonstrated by the diversity and breadth of the Victorian inventors. These inventors, like Edison, moved into new areas quickly as their creativity and interests changed. Edison could switch the use of his cognitive skills to the area of highest payback. Even the lower-profile meta-inventors, such as toolmaker Joseph Bramah, invented tools and pen nibs, as well as the beer keg pump. Their asset was in cognitive creation from streams of developing knowledge, not in specialized knowledge. They actually appeared to be bored staying too long on any particular project. Edison, the model of meta-inventors, loved to move in and out of projects.

Generalists, artists, and freethinkers such as Alexander Graham Bell helped to break mental paradigms that had slowed progress. The invention of the telephone, automobile, and airplane were all held up by mental paradigms.

In the case of the telephone, both inventors and science fiction writers were moving in different directions. Most saw a variation of the telegraph as the next communication advance. Even Jules Verne missed the telephone. Edison, working on a writing telegraph, was almost blinded by the paradigm. It took Bell's background as a voice coach to help break down that paradigm. Auto designs struggled first with the idea of steam locomotion without tracks, then the concept of the bicycle. Airplane design was inhibited by the mental image of the balloon, then by the idea of using flapping wings like birds. It is when a stream of innovation has a strong direction that generalists offer the best hope for a change of course. Specialists and bricklayers can advance theory and principles, but the Victorians showed the importance of the generalist in advancing technology. Bessemer, Edison, Nasmyth, Babbage, Bell, and Faraday were generalists who became known as meta-inventors. They had the ability to enter a new stream of information, summarize it and apply it in a new way. With the exception of Faraday, none of these meta-inventors was a scientist. Charles Murray's study in *Human Accomplishment* shows that the creativity burst of the Victorians was engineering-based not hard science–based. The hard sciences lend themselves today to system builders or brick laying.

If you accept Murray's premise that invention peaked around 1920, then it might be related to the lack of generalists. Research and science is dominated by corporate research, which moves down a path framed by set paradigms. The research centers are manned by specialists organized by specialties. Diversity is not favored and restricts the non-specialists' entry into other fields. With the exception of astronomy and computer science, amateurs are looked down on and rarely given serious consideration. Breaking a paradigm represents a slow and painful process, moving one brick at a time. Even boundaries between scientists and engineers are reinforced further, protecting old paradigms. These boundaries, restrictions, and reinforced paradigms were nonexistent in the Victorian Metallurgic Age.

The Victorian inventor represented the first engineer. He applied rather than advanced any scientific principle. Edison, the master inventor of all time, lacked the ability and desire to apply mathematics. He hired mathematicians and electrical scientists to help him with scientific theory. In many ways Edison favored DC current because of its simpler theoretical base. Bessemer similarly dealt little with pure science, preferring to leave science to the laboratory. Bessemer understood little of the steel-making chemistry but knew how to make steel. The true distinction between the scientist and engineer evolved during the Victorian Metallurgic Age. The scientist and engineer represent two very different skill sets, the engineer being visual rather than conceptual. The artist, as we have seen, can be considered a cousin of the engineer in that both use visual arts. The engineer must be creative to be successful, while the scientist must be intuitive. Scientists love mathematical representations and modeling, while engineers prefer physical modeling.

These meta-inventors were innovators using creative skills rather than specific training or research. Cognitive skills drive meta-inventions by analyzing and blending creative streams. My definition of a meta-invention differs from that of Charles Murray in *Human Accomplishment*. A meta-invention applies creativity to form technology into something practical. Meta-inventors such as Bessemer, Volta, Edison, Nasmyth, Oersted, and others were not technologists or specialized scientists but creative geniuses. They tended to favor engineering over science. They invented in diverse fields of interest. It was their creative problem solving that allowed them to move between fields unburdened by specific knowledge or credentials. Meta-inventors broke mental paradigms that restricted technological streams together. Bricklaying scientists had dominated the century before, while technical specialists dominated the century after. Murray classified Edison as a bricklayer, but this is far from the creative jumping around that characterized Edison's career. Edison added very little if any new scientific principles; he merely applied scientific principles in new ways. My definition of meta-invention stresses creativity, breaking mental paradigms, and bringing together streams of technology. Murray does not see any meta-inventions in technology, but finds them in the cognitive fields of science and philosophy. Murray does not see technology as cognitive in nature, but a result of applying science. Where Murray and I agree is that a meta-invention represents a mental breakthrough as much as a physical one. We agree also that a meta-invention "enables humans to do a class of new things" and that "transforming changes in practice and achievement" follow them. Meta-inventions are not necessarily in the headlines; in fact, they are often overshadowed. Nasmyth's steam forging hammer represented a meta-invention in that it allowed for a new class of ships, such as the *Great Eastern*. Oersted's magnetic coil pioneered the telegraph, telephone, and electromagnet. Whitney's tempered-steel patterns are rarely even mentioned in essays on interchangeable parts! Volta's battery caused a creative explosion in metallurgy and is the root of the whole field of telephony. Meta-inventions often are in history's attic. Engineers are in a better position to appreciate them than historians are. This can be seen in Samuel Morse's visit to Hans Oersted's European laboratory and comparing it to a pilgrimage to a religious shrine. Carnegie's visit in old age to William Kelly to thank the inventor is another example.

Back-to-back meta-inventions are truly rare even in today's modern science, but the Victorian years of 1800–1835 saw an explosion of meta-inventions and technology nodes. Galvani's frog leg experiments moved quickly to Volta's development of the battery. The battery is the most under-rated invention of all time, yet it tied together many creative streams into technological nodes. In 1810 the application of electroplating of gold and silver rocked the commodity market. The availability of electrical current led to the discovery of 47 metallic elements by Humphrey Davy in a 10-year period (1820–1829). By 1820, Hans Oersted, applying current through a wire coil, discovered the effect of electromagnetism, which opened up the whole field of telephony. The telegraph emerged by 1835. There is no com-parable short period of explosive creativity. Victorians had a deep sense of alacrity with the application of science. The rapid extension of scientific principles to engineering applications was possible because of the unique Victorian approach to creativity. Many times these Victorian engineers moved faster than the state of the technology they were applying and relied on science writers to keep the vision alive. Mary Shelley's story of Franken-stein for medical application of electric currents and Jules Verne's battery power on his *Nautilus* are examples. This propensity to move ahead of sci-entific logistics evolved from the economic motive. Edison, Bessemer, Kelly, Nasmyth, Oersted, Morse, Bell and the others openly admitted to this. In this respect the creative Victorians differed from the creative giants of the Renaissance, such as Da Vinci and Michelangelo, or the purists of the pre-vious century, such as Jefferson.

The Victorian meta-inventor was a unique breed in a unique environ-ment. Clearly the environment allowed him freely to apply his creativity. Today such inventors would be restricted by lack of credentials and knowl-edge. It is not that we lack genius today, but we lack an environment to fos-ter it. There are Bessemers and Edisons today, but it is unlikely they will ever surface in the sciences or engineering. Research and development has today been assigned to thecorporate efforts of highly educated specialists. While we have moved away from labor specialization in manufacturing, we have reinforced it in the sciences. Yet there have been a few explosive and innovative ("Victorian") breakthroughs in fields such as computers and robotics. These breakthroughs all reverted back to the Victorian model of streams of creativity versus corporate research. The Victorian model is not totally, lost as seen in companies such as 3M. The Victorian inventor achieved a type of self-actualization by combining passion, hope of rewards, and ambition. Carnegie's steel empire was built on the cornerstones of pas-sion, financial rewards, and ambition.

The lone Victorian inventor was favored economically versus the aggre-gate corporation. Westinghouse in building his creative empire realized that individual rights equaled the needed creative motivation for individuals to invent. Edison, who had benefited as a meta-inventor, built corporate research systems that had low productivity. Edison, and eventually General

Electric, took credit and controlled innovation at Menlo Park. Later Victorians created an environment that favored individual rewards for advances. Men such as Bessemer, Nasmyth and Edison applied their personal creativity to projects of the highest reward. There is at least some evidence that the decline of innovation in the last 70 years might be related to the rise of corporate or team research.[4] Later Victorians, such as Carnegie, Westinghouse, Schwab, and others, were concerned that this might happen. Even the development of management science during the Victorian Metallurgic Age focused on motivating the individual. Just as specialized assembly-line work can demoralize the worker, so can team research demoralize the inventor. Victorian writers such as Emerson and Thoreau immortalized the individual. Hero worship, while sometimes covering the underlying streams of creativity and knowledge, was the energy that kept the streams flowing. This is the irony of the Victorian image of the great engineer or scientist so loved by fiction writers and the reality of a stream of small advances. Being a bricklaying scientist or engineer lacks the inspiration of the breakthrough meta-inventor. Victorians understood creative streams but still wanted a hero to inspire them.

The individualistic emphasis of the Victorians not only influenced technology but also management science. The Victorian managers promoted an array of profit sharing, gain sharing, incentive plans, and awards to motivate employees. There was a love of competition at all levels, and a strong belief in individual competition. Corporate and team approaches were found to be inferior. The Edison approach, to treat creativity and innovation as a craft with a master craftsman, failed. Westinghouse and General Electric became the corporate models for individual motivation. Unfortunately both of these Victorian approaches have given way to a team model today. Another characteristic of the highly motivated employee innovator was the ability to rise quickly through the ranks based on his creativity. Carnegie and Westinghouse encouraged young engineers and mechanics through corporate advancement. Degrees and education were secondary to enthusiasm, creativity, and drive. Training and self-education were, however, encouraged.

The Victorian model of creativity is best envisioned as streams of creativity with confluences as the meta-inventions. The "great man" model clearly does not fit the explosion of similar inventions in the Victorian era. Take away any of the great Victorian inventors, such as Edison, and you simply delay the invention by a few years. In addition, the connective node model, while it makes for great history, doesn't fit either. Nodes are points that moved technology ahead by decades and were the foundations of meta-inventions to follow, such as the telegraph, electric motor and telegraph. Certainly the confluences of creative streams are nodes, but they are not revisionist. These nodes are the natural consequences of streams of creativity. Oersted's discovery of a relationship between electrical current and magnetism represents one of these Victorian nodes. Node inventors and scientists tended to be lost to history, but well-noted by the meta-inventors. Nielson's and Nasmyth's hot-blast developments in iron making were important nodes that led Bessemer and Kelly to the invention of pneumatic steel making. Node inventors are more than brick-laying scientists in that they bring streams together in novel ways, like Volta taking the animal/metal current experiments of Galvani and developing the electrical battery. They apply science in a new vein or manner. Their inventions often had little economical value but were inspirational to other inventors.

Tracing nodes back from a particular invention is a great way to write history, but it is not causal proof. Science writer James Burke used this technique in his popular TV series and book *Connections*. The history of Victorian invention shows nodes, but causal linkage is difficult as you move back in history. Nodes (or maybe a better term is confluences) brought streams together, but their linkage was not necessary to the appearance of significant inventions. These nodes, however, allowed for the faster development of many inventions. Looking back, the connections appear, but this is a revisionist view. Maybe just as important was the strength of the creative streams, the strength being a result of motivation and inspiration. Strength is reflected in Emerson's famous quote: "Invention breeds invention." Nodes, like the meta-inventions, are the result of the creative streams. In a creative environment such as the Victorian Metallurgic Age, invention and creativity can be explosive, making the nodes almost secondary events. Nodes are more significant in brick-laying environments such as this century and the 18th century. Volta's battery allowed Faraday to discover 47 new metals by using electrical current to reduce them from their ores. Volta's work and even Galvani's a few years earlier were important nodes.

The best example of nodes and creative streams is the sewing machine. The sewing machine required standardized machined parts brought together in a unique application. You could link nodes in the development of machined parts or advances in the automated textile equipment in the 1700s, but the impact of creativity would be lost. From today's perspective, it would seem logical to link Richard Arkwright's 1770 automated spinning machine as a node in the development of the sewing machine. A closer view, however, might be to look at the streams of machined parts and machine tools being brought together by a mechanic, Elias Howe. Howe was not inhibited by the idea of imitating the hand-sewing process. The textile industry had been following that paradigm in automating hand processes such as knitting. Howe stood back, developing a stitch not like the hand stitch, but designed to fit the machinery. He invented from the perspective of the mechanic, not the sewer. This is the characteristic of a meta-inventor to bring streams together in a new way. Howe's device can be considered a node because it brought mechanical sewing into new applications, such as bookbinding and shoemaking. Of course, the commercial victory did not go to Howe but to others. Ultimately it was Isaac Singer in 1851 who patented the commercially successful sewing machine.

Creativity belonged not just to the inventor but to the industrial managers as well. Creative managers such as the "boys of Braddock" found innovative ways to apply technology and motivate workers. The industrial manager, like the inventor of the Metallurgic Age, became heroes of the youth. Basic industry and industrial towns drew ambitious youths to seek their fortune. "The myth of the steel millionaire also seduced the people of the steel towns," William Serrin says. "Carnegie, Frick, Schwab, Julian Kennedy, Alva Dinkey, William E. Corey, and his brother A. A. Corey, William Dickson, A. R. Hunt [the "boys of Braddock"]—they and others started poor and became rich, important men. They are regarded as representatives of that quintessential American hero, the self-made man, and were honored by the townspeople. Many a mill-town boy, especially in the early days, went into the mill with the hope of making himself rich, a big man in steel."[5] This was the power and seduction of the Metallurgic Age. Industry offered the hope to move out of poverty, and while many argue the actual reality, the careers of many show it was more possible then than now. These young recruits came to work with enthusiasm, bubbling creativity, a burning desire to succeed, and an optimism that favored innovation. Men like Bill Jones and Julian Kennedy found their creative powers ignited by the dream of becoming industry leaders. The very possibility of achieving the dream pulls creativity out of the recesses of the organization.

Are there meta-inventors today? Certainly Steve Jobs of Apple and Bill Gates of Microsoft would fit the Victorian mold for meta-inventors. Jobs understood that invention was a stream of creativity, not the brainchild of any individual. The personal computer brought together several creative streams. Jobs has boldly said, "Good artists copy; great artists steal," a corporate mantra. He even flew a pirate flag at the corporate headquarters of Apple to highlight this point. Jobs's honesty goes directly to the nature of Victorian creativity and invention. "Copying" and "stealing" were central to Victorian advances. Certainly this could be said for Bessemer and Kelly, Bell and Gray, and so many others. Jobs, like those inventors, was the creator that pulled several streams of technology together. Like many of the successful Victorian inventors, he lost the patent battles but won the commercial battle. In fewer then 20 years, Jobs's and Gates's success erased the original inventors of Xerox from the history books. Gates, even more so than Jobs, tied together streams of technology to create the "graphical interface." Jobs and Gates of course, invented nothing, but they created an industry much like Bessemer, Morse, and Bell.

Another area in which we might be seeing a Victorian-style renaissance of creativity is robotics. To a large degree, the science of robotics is still in its infancy. Certainly robots in welding and painting manufacturing uses have been around for 20 years, using earlier advances in automation, but we are now just approaching the robotic humanoid. Many of the Victorian factors are in place for a breakthrough in humanoid robotics. Here we see streams of technology in remote control, computer, optical sensing, and wireless transmission coming together. The science fiction writers of the 1950s, much like their Victorian inspiration, defined the nature of robots. Hobbyists have recently been spurred by rewards and conferences to new advances. The military has even encouraged the entrance of hobbyists in the field through contests and prizes. Movies and books are again focusing on robots versus fantasy. We see the kind of passion and intermixing of scientists, engineers, and hobbyists. The history of Victorian invention would suggest a breakthrough in humanoid robotics by the end of the decade.

By contrast, we can look at the more corporate and controlled streams of science today. Hobbyists are blocked out, and science fiction writers offer no visions. Progress in these areas is slow but steady. These institutional and corporate scientists and engineers are bricklayers. Victorians sowed the seeds of this modern leveling off of creativity as well. Successful Victorian industrialists such as Charles Schwab gave money to build university departments of engineering, metallurgy, and chemistry. Emphasis on technical degrees and specialization helped choke the roots of Victorian creativity. Business schools and management degrees blocked the Victorian career routes from the factory floor to the boardroom.

Chapter Notes

Chapter 1

1. Herbert Casson, *The Romance of Steel*.
2. Casson, 134.
3. Asa Briggs, *Victorian Things* 185.
4. James Dugan, *The Great Iron Ship* 1.
5. Charles Murray, *Human Accomplishment: The Pursuit of Excellence in Arts and Sciences 800 B.C. to 1950*, 163–170.
6. Isaac Asimov, *A Short History of Chemistry*.
7. Carroll Pursell, ed. *Technology in America*, 63.
8. Ralph Waldo Emerson, *The Works of Ralph Waldo Emerson*, 228.
9. Clarice Swisher, *Victorian England*, 12.
10. Charles Murray, *Human Accomplishment*, 198–203.
11. "The American Mechanic," *New York Evening Telegram*, April 25, 1878.
12. Arthur B. Evans, *Jules Verne Rediscovered*, 118.
13. Henry Petroski, *Remaking The World*, 44.
14. Murray, 437.
15. Walter E. Houghton, *The Victorian Frame of Mind*, 305.
16. Jared Diamond, *Guns, Germs, and Steel*, 244.
17. Robert Breeden, ed. *Those Inventive Americans*, 90.
18. Sale quoted in Swisher, 120.
19. *Bulletin of the American Iron and Steel Association*, April 10, 1901, 52.
20. Briggs, 19.
21. James Burke, *The Pinball Effect*, 118.
22. Donald Cardwell, *Wheels, Clocks, and Rockets*, 237.
23. Witold Rybczynski, *One Good Turn*, 110.
24. Fredric Harrison, *Autobiographic Memoirs*.

Chapter 2

1. Walter E. Houghton, *The Victorian Frame of Mind*, 40.
2. Hermione Hobhouse, *Prince Albert: His Life and Work*, 90.
3. Hobhouse, 91.
4. Donald Cardwell, *Wheels, Clocks, and Rockets*, 286.
5. Clarice Swisher, *Victorian England*, 78.
6. Swisher, 80.
7. Henry Petroski, *Remaking The World*, 173.
8. "The American Abroad," *New York Daily Tribune*, August 18, 1889.
9. Murat Halstead, "Electricity at the Fair," *Cosmopolitan*, September 15, 1893.
10. Reid Badger, *The Great American Fair*, 113.

Chapter 3

1. Arthur B. Evans, *Jules Verne Rediscovered*, 25.
2. Robert V. Bruce, *Lincoln and the Tools of War*, 176.
3. James Dugan, *The Great Iron Ship*, 2.
4. Henry Petroski, *Remaking the World*, 131.
5. Paul Israel, *Edison*, 364.
6. Israel, 366.
7. John Brunner, ed. *The Science Fiction Stories of Rudyard Kipling*, 14.

Chapter 4

1. Arthur Cecil Bining, *Pennsylvania Iron Manufacture in the Eighteenth Century*, 63–64.
2. James Howard Bridge, *The Inside History of the Carnegie Steel Company*, 46.
3. Peter Temin, *Iron and Steel in Nineteenth Century America*, 26.
4. Bridge, 142.
5. Herbert Casson, *The Romance of Steel*, 88.
6. Sara Ruth and Emily Watson, *Famous Engineers*, 97.
7. Douglas Fisher, *The Epic of Steel*, 104.

Chapter 5

1. Robert Hessen, *Steel Titan: The Life of Charles M. Schwab*.
2. John Newton Boucher, *William Kelly: A True History of the So-Called Bessemer Process*.
3. Norbert Muhlen, *The Incredible Krupps*, 37.
4. William T. Jeans, *The Creators of the Age of Steel*.
5. Stewart Holbrook, *Iron Brew*, 3.
6. Boucher, 115.
7. Lewis C. Walkinshaw, *Annals of Southwestern Pennsylvania*, 136.
8. David McCullough, *The Johnstown Flood*, 23.
9. Boucher, 57.
10. Boucher, 10.
11. Herbert Casson, *The Romance of* Steel, 14.
12. Boucher, Stewart Holbrook, *Iron Brew*.
13. William T. Jeans, 38–39.
14. Deborah Cadbury, *Terrible Lizard*, 241.
15. James Howard Bridge, *The Inside History of the Carnegie Steel Company*.
16. H. Hearder, *Europe in the Nineteenth Century*, 81.
17. Peter Krass, *Carnegie*, 223.
18. Krass, 269.
19. John Emsley, *Nature's Building Blocks*, 503.

Chapter 6

1. Robert V. Bruce, Lincoln and the Tools of War, 100.

2. Fredrick Wilkinson, *The Illustrated Guns and Rifles*, 101.
3. William Manchester, *The Arms of the Krupp*, 112.
4. Bruce, 87.
5. Alden R. Carter, *Battle of the Ironclads*, 20.

Chapter 7

1. J. G. Pangborn, *The Golden Age of the Steam Locomotive*, 19.
2. Stephen Goodale, *Chronology of Iron and Steel*, 174.
3. Peter Krass, *Carnegie,*, 88.
4. Rhoda Blumberg, *Full Steam Ahead*, 49.
5. T. K. Derry and Trevor Williams, *A Short History of Technology*, 459.
6. Blumberg, 140.
7. Henry Petroski, *Engineers of Dream*, 98.

Chapter 8

1. S. I. Venetsky, *Tales about Metals*, 29.
2. Venetsky, 31.
3. Edward B. Tracy, *The World of Aluminum*, 39.
4. Junius David Edwards, Francis C. Frary and Zay Jeffries, *The Aluminum Industry*, Vol. 1, 27.
5. Junius David Edwards, *A Captain in Industry*, 72.

Chapter 9

1. Miranda Seymour, *Mary Shelley*, 43.
2. Percy Dunsheath, *Giants of Electricity*, 70.
3. Kenneth Silverman, *Lighting Man*, 347.
4. Silverman, 151.
5. Silverman, 159.
6. Silverman, 326.
7. John Steele Gordon, *A Thread across the Ocean*, 172.
8. Ronald Clark, *Edison*, 57.
9. Clark, 57.
10. Carroll W. Pursell, ed. *Technology in America*, 112.
11. Asa Briggs, *Victorian Things*, 384.

Chapter 10

1. William Manchester, *The Arms of the Krupp*, 206.
2. T. K. Derry and Trevor Williams, 358.
3. John Emsley, *Nature's Building Blocks*, 257.
4. S. I. Venetsky, *Tales about Metals*, 190.

Chapter 11

1. Henry Petroski, *The Pencil*, 22.
2. Walter Harding, *The Days of Henry Thoreau*, 56.
3. Petroski, *The Pencil*, 169.
4. Carroll Pursell, ed., *Technology in America*, 48.
5. Pursell, 55.
6. Henry Metcalfe, *The Cost of Manufacturers and the Administration of Workshops*, 15.
7. Witold Rybczynski, *One Good Turn*, 76.

Chapter 12

1. Claude George, *The History of Management Thought*, 58.
2. Erick Roll, *An Early Experiment in Industrial Organization*, 222.
3. Harwood Merrill, *Classics in Management*, 33.
4. Charles Babbage, *On the Economy of Machinery and Manufacturers*, (London: Charles Knight, 1832), 132.
5. W. S. Jevons, *The Theory of Political Economy*, 204.
6. "The Building of American Locomotives," *Scientific American*, August 23, 1902.
7. Gerald Eggert, *Steel Masters and Labor Reform*, 108.
8. Eggert, 161.

Chapter 13

1. T. K. Derry and Trevor Williams, *A Short History of Technology*, 513.
2. Robert Conot, *A Streak of Luck*, 118.
3. Robert Silverberg, *Light for the World*, 93.
4. Conot, 472.

5. Conot, xvii.
6. Richard Shelton, Sidney Withington, Arthur Darling, and Fredrick Kilgour, *Engineering in History*, 356.
7. Conot, 258.
8. Carroll Pursell, ed., *Technology in America*, 122.
9. Silverberg, 97.
10. Silverberg, 150.
11. Paul Israel, *Edison*, 336.
12. John Winthrop Hammond, *Men and Volts*, 203.

Chapter 14

1. Francis E. Leupp, *George Westinghouse: His Life and Achievements*, 44.
2. Leupp, 48.
3. Robert Silverberg, *Light for the World*, 230.
4. Silverberg, 232.
5. Jill Jones, *Empires of Light*, 179.
6. Silverburg, 240.
7. I. E. Levine, *Inventive Wizard*, 158.
8. Silverburg, 260.
9. Barbara Ravage, *George Westinghouse*, 95.
10. Leupp, 245.

Chapter 15

1. John Rae, *American Automobile Industry*, 10.
2. T. K. Derry and Trevor Williams, *A Short History of Technology*, 601.
3. Richard Scharchburg, *Carriages without Horses*, 6.
4. Horst Hardenburg, *Samuel Morey and His Atmospheric Engine*.
5. Scharchburg, 112.
6. Daniel A. Wren, *History of Management Thought*, 263.

Epilogue

1. Timoty Ferris, *Seeing in the Dark*, 159.
2. Petroski, *Remaking the World*, 106.
3. Richard Florida, *The Rise of the Creative Class*, 31.
4. Charles Murray, 436.
5. William Serrin, *Homestead*, 53.

Bibliography

Asimov, Isaac. *A Short History of Chemistry*. New York: Anchor Books, 1965.

Babbage, Charles. *On the Economy of Machinery and Manufacturers*. London: Charles Knight, 1832.

Badger, Reid. *The Great American Fair*. Chicago: Nelson Hall, 1979.

Bining, Arthur Cecil. *Pennsylvania Iron Manufacture in the Eighteenth Century*. Harrisburg: Pennsylvania Historical Commission, 1979.

Blumberg, Rhoda. *Full Steam Ahead*. Washington DC.: National Geographic Society, 1996.

Boucher, John. *William Kelly: A True History of the So-Called Bessemer Process*. Greensburg, Pennsylvania: privately published, 1924.

Breeden, Robert, ed. *Those Inventive Americans*. Washington DC: National Geographic Society, 1971.

Bridge, James Howard. *The Inside History of the Carnegie Steel Company*. New York: Aldine Books, 1903.

Briggs, Asa. *Victorian Things*. Chicago: University of Chicago Press, 1988.

Bruce, Robert V. *Lincoln and the Tools of War*. Urbana: University of Illinois, 1989.

Brunner, John, ed. *The Science Fiction Stories of Rudyard Kipling*. New York: Carol Publishing Group, 1994.

Burke, James. *The Pinball Effect*. Boston: Little, Brown, and Company, 1996.

Cadbury, Deborah. *Terrible Lizard*. New York: Henry Holt and Company, 2000.

Cardwell, Donald. *Wheels, Clocks, and Rockets*. New York: W. W. Norton, 1994.

Carter, Alden R. *Battle of the Ironclads*. New York: Franklin Watts, 1993.

Casson, Herbert. *The Romance of Steel*. New York: A. S. Barnes & Company, 1907.

Clark, Ronald. *Edison*. New York: G. P. Putman's Sons, 1977.

Claude, George. *The History of Management Thought*. Englewood: Prentice-Hall, 1968.

Conot, Robert. *A Streak of Luck*. New York: Seaview Books, 1979.

Derry, T. K., and Trevor Williams. *A Short History of Technology*. New York: Dover Publications, 1960.

Diamond, Jared. *Guns, Germs, and Steel*. New York: W. W. Norton & Company, 1997.

Dugan, James. *The Great Iron Ship*. New York: Harper & Sons, 1953.

Dunsheath, Percy. *Giants of Electricity*. New York: Thomas Crowell Company, 1967.

Edwards, Junius David. *A Captain in Industry*. New York: privately printed, 1957.

Edwards, Junius David, Francis C. Frary, and Zay Jeffries. *The Aluminum Industry*, Vol 1. New York: McGraw-Hill, 1930.

Eggert, Gerald. *Steel Masters and Labor Reform*. Pittsburgh: University of Pittsburgh Press, 1981.

Emerson, Ralph Waldo. *The Works of Ralph Waldo Emerson*. New York: Tudor Publishing.

Emsley, John. *Nature's Building Blocks*. Oxford: Oxford Press, 2001.

Erick, Roll. *An Early Experiment in Industrial Organization.* London: Longmans, Green & Co., 1930.

Evans, Arthur D. *Jules Verne Rediscovered.* New York: Greenwood, 1988.

Ferris, Timothy. *Seeing in the Dark.* New York: Simon & Schuster, 2002.

Fisher, Douglas. *The Epic of Steel.* New York: Harper & Row, 1963.

Florida, Richard. *The Rise of the Creative Class.* New York: Basic Books, 2002.

George, Claude. *The History of Management Thought.* Englewood: Prentice-Hall, 1968.

Goodale, Stephen. *Chronology of Iron and Steel.* Cleveland: Penton Publishing, 1931.

Gordon, John Steele. *A Thread across the Ocean.* New York: Walker & Company, 2002.

Hammond, John Winthrop. *Men and Volts.* New York: Lippincott Company, 1941.

Hardenburg, Horst. *Samuel Morey and His Atmospheric Engine.* Warrendale, PA.: Society of Automotive Engineers, 1992.

Harding, Walter. *The Days of Henry Thoreau.* New York: Dover Publications, 1962.

Hardwood, Merrill, ed. *Classics in Management.* New York: American Management Association, 1960.

Harrison, Frederic. *Autobiographic Memoirs,* 2 vols. London, 1911.

Hearder, H. *Europe in the Nineteenth Century.* London: Longmans, 1966.

Hessen, Robert. *Steel Titan: The Life of Charles M. Schwab.* New York: Oxford Press, 1975.

Hobhouse, Hermione. *Prince Albert: His Life and Work.* London: Garden House, 1983.

Holbrook, Stewart. *Iron Brew.* New York: The Macmillan Company, 1939.

Houghton, Walter E. *The Victorian Frame of Mind.* New Haven: Yale University Press, 1957.

Israel, Paul. *Edison.* New York: John Wiley & Sons, 1998.

Jeans, William T. *The Creators of the Age of Steel.* New York: Charles Scribner's Sons, 1884.

Jevons, W. S. *The Theory of Political Economy.* New York: Macmillan and Company, 1888.

Jones, Jill. *Empires of Light.* New York: Random House, 2003.

Krass, Peter. *Carnegie.* Hoboken: John Wiley & Sons, 2002.

Leupp, Francis. *George Westinghouse: His Life and Achievements.* Boston: Little, Brown, and Company, 1918.

Levine, I. E. *Inventive Wizard.* New York: Julian Messner, 1962.

Manchester, William. *The Arms of the Krupp.* Boston: Little, Brown and Company, 1964.

McCullough, David. *The Johnstown Flood.* New York: Touchstone Books, 1986.

Merrill, Harwood. *Classics in Management.* New York: American Management Association, 1960.

Metcalfe, Henry. *The Cost of Manufacturers and the Administration of Workshops.* New York: John Wiley & Sons, 1885.

Muhlen, Norbert. *The Incredible Krupps.* New York: Award Books, 1969.

Murray, Charles. *Human Accomplishment.* New York: HarperCollins, 2003.

Pangborn, J. G. *The Golden Age of the Steam Locomotive.* New York: Dover Publications, 1894.

Petroski, Henry. *Engineers of Dreams.* New York: Vintage Press, 1996.

_____. *The Pencil.* New York: Alfred A. Knopf, 1989.

_____. *Remaking the World.* New York: Vintage Books, 1997.

Pursell, Carroll, ed. *Technology in America.* Cambridge: The MIT Press, 1981.

Rae, John. *American Automobile Industry.* Boston: Twayne Publishers, 1984.

Ravage, Barbara. *George Washington.* Austin: Raintree Steck-Vaughn, 1997.

Roll, Erick. *An Early Experiment in Industrial Organization.* London: Longmans, Green & Co., 1930.

Ruth, Sara, and Emily Watson. *Famous Engineers.* New York: Dodd, Mead & Company, 1950.

Rybczynski, Witold. *One Good Turn.* New York: Scribner, 2000.

Scharchburg, Richard. *Carriages without Horses.* Warrendale: Society of Automotive Engineers, 1993.

Serrin, William. *Homestead.* New York: Times Books, 1992.

Seymour, Miranda. *Mary Shelley*. New York: Grove Press, 2000.
Shelton, Richard, Sidney Withington, Arthur Darling, and Fredrick Kilgour. *Engineering in History*. New York: Dover Publications, 1990.
Silverberg, Robert. *Light for the World*. Princeton: D. Van Nostrand, 1967.
Silverman, Kenneth. *Lighting Man*. New York: Alfred Knopf, 2003.
Skrabec, Quentin. *The Boys of Braddock*. Westminster: Heritage Books, 2004.
Swisher, Clarice. *Victorian England*. San Diego: Greenhaven Press, 2000.
Temin, Peter. *Iron and Steel in Nineteenth Century America*. Cambridge: MIT Press, 1964.
Tracy, Edward B. *The World of Aluminum*. New York: Dodd, Mead & Company, 1967.
Venetsky, S. I. *Tales about Metals*. Moscow: Mir Publishers, 1978.
Walkinshaw, Lewis C. *Annals of Southwestern Pennsylvania*. New York: Lewis Historical Publishing, 1939.
Warren, Kenneth. *Triumphant Capitalism*. Pittsburgh: University of Pittsburgh Press, 1996.
Wilkinson, Fredrick. *The Illustrated Guns and Rifles*. London: Optimum Publishing, 1979.
Wren, Daniel A. *History of Management Thought*. New York: John Wiley & Sons, 2004.

Index

AC current 208–222
Air brakes 210–211
Aldini, Giovanni 133
Alligator 43–46
Alloying 148–149
Aluminum 31, 41–43, 120–129, 130–134, 220
Aluminum bronze 128
Aluminum car body 129
Aluminum foil 128
Aluminum in aircraft 129
Amateurism 251–253
American Institute of Mining, Metallurgical, and Petroleum Engineers 32
American Society of Mechanical Engineers 10, 189–190
Ampère, André-Marie 135, 136
Anode 131
Arc lighting 194
Arch bridges 112–113
Armor 151–152, 155, 157–158
Armstrong, William 101–102
Atchison, Topeka, and Santa Fe Railroad 107
Atlantic cable 141–144
Automobile 223–228

Babbage, Charles 29, 50, 173–174, 182–183
Baggaley, Ralph 210–211
Baker, Benjamin 114
Bain, Alexander 141–142
Baldwin Locomotive 184–185
Balloons 98–100
Baltimore and Ohio Railroad 103–105, 139
Batchelor, Charles 200
Batteries 130–135
Baur, Julius 149
Bauxite 124–125
Bayer, Karl Joseph 120–126

Bell, Alexander 35, 143–144
Bellamy, Edward 50
Benz, Karl 226–227
Berlin Exhibition of 1881 216
Bessemer, Henry 15, 63–83, 101, 160
Bessemer Process 63–83, 105–106, 156, 156–157
Birmingham 3
Blackwell, Fredrick 141
Blast furnace 55–60
Boehm, Ludwig 200
Bogardus, James 57
Bouch, Thomas 113
Boulton, Matthew 180–181
Boys of Braddock 22, 63, 187–188, 190, 238
Bramah, Joseph 168–170
Brass 132, 160
Brayton, George 225–226
Breech loading 86–88, 95–97, 149, 175
Britannia Bridge 60–61
British Association for the Advancement of Science 10, 72
Bronze 89–90, 96, 132, 148, 160
Brooklyn Bridge 7, 61, 83, 115–116
Brown, Harold 217–129
Brugnatelli, Luigi 158
Brunel, Isambard Kingdom 17, 46, 60–62, 116–117, 147, 161
Brunel, Marc 116–117, 169
Brush, Charles 195–198
Bunsen, Robert Wilhelm 132

Cadmium 132
Cannon 9, 29, 43–44, 65–66, 84–102, 151–152, 175
Cantilever bridges 114–115
Cambria Iron Company 67–70, 74
Carbon in steel 51–53, 62–67, 72–74, 132
Cardwell, Donald 28, 30

Carnegie, Andrew 8, 14, 58, 63–65, 73–
    75, 80–83, 105–106
Carnegie, Tom 76–77, 105
Carnegie Steel 58, 73–76
Casey, Thomas Lincoln 123–124
Cast iron 7–8, 29, 51–62, 75, 103–106,
    112–113, 147
Cast iron pipes 57
Cast iron wheels 107–108
Catalan forge 64–65
Centennial Exhibition of 1876 9, 31–33,
    123, 145, 193–194, 225
Chappe, Claude 137
Charcoal Furnaces 51–56
Chernyshevsky, Nikolai 42
Chromium in steel 83, 102, 105, 132, 148,
    159
Civil War 85–102, 166–168
Clement, Joseph 173–174
Coalbrookdale Bridge 112
Coke 53–60
Coleman, William 76
Colt, Samuel 175–176
Columbiad cannon 89–92
Columbian Exposition of 1893 34–35, 37,
    153, 219, 226
Comte, Lazare 163
Conan Doyle, Arthur 13, 48
Conot, Robert 197–198
Cooke, William 37
Cooper, Peter 61–62, 89, 103–105, 142–
    143
Cooper, Theodore 114–115
Copper 130–136, 142–143, 148, 150, 203–
    205, 216–217, 220
Copper-nickel 150
Corrosion 159–160
Cort, Henry 64–65
Creativity 233–239
Crucible process 65–67, 69, 156
Crucible steel 83
Crystal Palace 7, 9, 20, 25–27, 231
Cugnot, Nicholas 223

Dahlgren, Captain John 85–86, 90–91,
    166–167
Dahlgren guns 100
Daimler, Gottlieb 226–227
Damascus steel 154–155
Darby, Abraham 111–112
Darwin, Charles 2
Davy, Humphrey 21, 134, 147–148, 194
Deville, Henri Ètienne Sainte-Clare 20,
    42, 120, 121–122, 123, 125, 126, 129

Diamond, Jared 23
Dickens, Charles 108, 122
Dickson, William 191–192
Dixon, Joseph 163–164
Draper, Henry 197
Dumas, Alexandre 33–34
Duralumin 129
Durand, Peter 59–60
Duryea, Frank 226–228

Eads, James 44, 82–83, 99–101, 113
Eads Bridge 61, 82, 113
Edgar Thomson steel works 68–70, 74,
    76–78, 80–81, 105, 187–189
Edison, Thomas 13, 34, 49, 144–146, 163,
    196–203, 208–222
Edison bulb 219–220
Eiffel, Gustave 33, 149, 159
Eiffel Tower 7, 9, 33–34, 51
Einstein, Albert 232–233
Electric chair 217–219
Electric furnaces 121–127
Electricity 130–146
Electroplating 156–159
Eliot, George 13
Emerson, Ralph Waldo 18, 30, 116
Ericsson, John 100, 104
Essen 3

Faber, A.W. 163
Fabergé, Carl 160
Facsimile 141
Faraday, Michael 4, 136–138, 212–213
Farmer, Moses 197
Ferris, Washington 35
Ferrochromium 127, 148
Field, Cyrus 141–142
Filaments 200–203
Firth of Forth Bridge 83, 113–114
Flad, Henry 112
Ford, Henry 155
Fort Pitt Foundry 91–94
Forth Bridge 83
Fowler, John 83, 114
Franco-Prussian War 38, 43, 89, 96
Franklin, Benjamin 130, 134
Franklin, John 59
Frick, Henry Clay 82–83, 128
Frishmuth, William 20, 123–124
Froude, William 118
Fulton, Robert 44

Gale, Leonard 138
Galvani, Luigi 130–132

Galvanic cell 2, 131–135
Galvanic protection 158–160
Galvanized steel 116, 142
Galvanizing 83–84, 142
Gates, Bill 12
Gatling, Richard 89
Gatling guns 88–90
Gaulard, John 213
Gaulard-Gibbs transformer 212–213
Gauss, Karl 137
Gayley, James 80
General Electric 220–221
German silver 159
Gibbs, John 213–214
Gilchrist-Thomas Process 78–79
Gilded Age 1, 7
Gilding 158–159
Gillott, Joseph 176–177
Globel, Henricg 201–202
Gold 21, 30, 132, 158–160
Goodyear, Charles 28
Goubet II 11
Gramme dynamo 195–197
Gratzer, Walter 11
Gray, Elisha 18, 143–144
Gray iron 106–107
Great Britain 117
Great Eastern 8, 17, 44, 46, 117, 142–143, 161
Great Exhibition of 1851 5, 9, 16, 20–21, 25, 29–31, 101, 161, 166, 168, 170, 204
Great Western 117
Great Western Railway Company 117
Grove battery 139

Hadfield, Robert 149–150, 153–154
Hadfield alloy 153–154
Hall, Charles 36, 120, 125
Hall-Heroult Process 125
Hardening of steel 148–152
Harvey, Hayward 152–153
Harvey process 152–153
Henry, Joseph 97–99, 134–136
Hero worship 17–18, 47
Heroult, Paul 122–126
Hetzel, Pierre 38–39, 47
Hocking Hills iron district 56
Holland 118
Holland, John 118–119
Holly, Alexander 32, 68–73, 79
Houghton, Walter 17, 26, 47
Houston, Edwin 206–207
Hunley 44–45, 99, 118

Hunt, Alfred 36, 126–128
Huntsman, Benjamin 65–66
Incandescent lamps 195–205, 214
International Nickel 153–154
Iridium 176–177, 197
Iron furnaces 51–55

Jablochkoff, Paul 194–197
Jones, Captain Bill 80–81, 186–189
Jones, Thomas P. 11
Journal of the Franklin Institute 11
Judah, Theodore 109

Kelly, William 15, 63–82
Keystone Bridge Company 115
Kipling, Rudyard 50, 158
Krupp, Alfred 25, 32, 36, 65–66, 93–96, 98–100, 152
Krupp works (Germany)/Krupp steel 9, 29, 31, 66, 77–78, 90, 93, 95, 96, 107, 148, 152, 156–157

Lead 131–134
Lee, Roswell 165
Lenoir, Etienne 225
Lincoln, Abraham 5, 44–45, 85, 95–97
Louis XVI 111–112

Machine guns 88–90
Machining 165–172
Magnesium 132–134
Malleable iron 56–59, 63–64, 70–71
Mallory, Stephen 99–100
Manchester 3
Manchester and Liverpool Railroad 104–105
Manganese 74–75, 102, 149
Marcus, Siegfried 225–226
Martens, Adolph 178
Mason, Sir Josiah 21, 176–177
Maudslay, Henry 169–175, 179–180
McCallum, Daniel 184–185
Menai Straits Bridge 60, 112
Menlo Park 198–200
Mercury 158
Merrimac 9–102
Meta-invention 2–4, 9, 19
Metcalfe, Henry 167–168
Midvale Steel 189–192
Molybdenum 155–156
Monell, Ambrose 153–154
Monitor 100, 167
Monthly Engine Reporter 12
Morey, Samuel 224–226

Morgan, J.P. 203–205
Morrell, Daniel 68–69
Morse, Samuel 18, 137–140
Mount Cenis 109–110, 210
Murray, Charles 3, 9, 18, 22
Mushet, Robert 74–75, 149–150, 154
Mushet making 164–166
Mushet's self-hardening steel 154–155

Napoleon 20, 44, 59, 130
Napoleon III 5, 31, 42, 89, 121–122, 141
Napoleon 12-pound cannon 89–91
Nasmyth, James 16, 71–73, 147–148, 161–162, 169–172, 179
Nature 11
New Harmony 182
New Lanark 181–182
Newcomen steam engine 23
Niagara Falls power station 219–220
Niagara Falls railroad bridge 61
Niagara Gorge Bridge 115
Nicholas I 137
Nickel 132, 150–152
Nickel-silver 150
Nickel steels 151–152
Niobium 127
Nobel, Alfred 110
Noble metals 176

Oersted, Hans 4, 121, 134–138
Ohm, George 202
Open-hearth steel 114–115, 150, 156–157
Osmium 176–177
Owen, Robert 181–182, 221–222

Pacific Railroad Act of 1862 109
Paktong 150
Paris International Exhibition of 1889 9, 19, 33–34, 51, 150, 154, 174, 178, 197, 204
Parrott, Robert 90–91
Parrott cannon 90–92
Parry, William Edward 59
Patent office 14, 70, 144
Paxton, Joseph 27
Pearl Street Station 217–219
Pen making 169–170, 176–177
Pencil making 163–164
Pennsylvania Railroad 67–70, 74, 182, 210
Perronet, Jean 111
Pfeiffer, Charles 112
Philips, Henry 76–77
Phosphorus 74, 77–78, 156
Pig iron 53–54, 63–68

Pittsburgh 3, 36, 108
Pittsburgh Reduction Company 36, 127
Pittsburgh Regional Industrial Exposition 36
Pixii, Hippolyte 194
Platinum 132, 139, 160, 197, 201
Pliny the Elder 120
Plunger 118
Popular Science 12
Porter, Rufus 108–109
Pratt, Francis 175–176
Prince Albert 5, 20, 25–28, 147
Puddling furnace 53–54, 63–65
Pullman Palace Company 111

Queensboro Bridge 153

Railroad rails 8, 67–70, 73–76, 103–105
Railroad wheels 106–107
Red lead 159
Regulation of Railways Act of 1871 113
Rice, Edwin 206–207
Riley, James 151–152
Roberts, Richard 170
Rodman, Major Thomas 90–91, 167
Rodman cannon 90–94, 167–168
Roebling, John A. 60–61, 83, 115–116
Roebling, Washington 115

Sacramento Valley Railroad 109
St. Gotthard Tunnel 110
Saint-Simonianism 17
Sauveur, Albert 178
Schilling, Paul 137
Schneider, Charles 114
Schwab, Charles 5, 13, 20, 69–70, 189–191
Scientific American 11–12
Screw-cutting machine 168–170
Screw propeller 117–118
Senarens, Luis 49
Shelley, Mary 133
Siemens Electric 34
Siemens-Martin process 156
Silver 131, 158–159, 220
Sixth Street Bridge 61
Sodium 134
Soho Foundry 180–181, 194
Sommeiler, Germain 109
Spencer, Christopher 88–89
Stainless steel 159
Statue of Liberty 159
Steam drills 109–110
Steam hammer 171–173

Steam locomotive 7, 102–106, 170–172
Steel 8–9, 43–44
Steel production 8
Steel rails 65–68
Steel wire 83
Stephenson, Robert 60–61, 104–107
Stoddard, William 86
Sturgeon, William 135–136
Submarine 43–45, 97, 99–102, 118–119
Sulfur 156
Summit Tunnel 110
Suspension bridges 60–61, 114–115
Swan, Joseph 196–197, 199
Sweet, John Edson 32
Sword making 6, 154–155

Tantalum 127
Tay Bridge 113–114
Taylor, Fredrick 5, 23, 177, 185–186, 189
Telegraph 132–140
Telephone 144–146
Telford, Thomas 60, 112–113
Tesla, Nikola 214–216, 221
Thomas, Percy 78–79
Thomas, Sidney Gilchrist 78–79
Thomson, Edgar 105–106
Thomson, Elihu 206–207
Thomson-Houston Company 206–207
Thoreau, Henry 5, 22, 163–164
Tin 132
Tin cans 59–60
Tin plating 59–60
Titanium 132
Tool steels 156–158, 176–178
Towne, Henry Robinson 177–178, 189–190
Transcontinental Railroad 109
Tredegar Works 91, 100
Trevithick, Richard 103–104
Tungsten 153–156, 178, 206–207
Twain, Mark 8, 88

Union Pacific Railroad 109
Union Switch and Signal Company 211–212
Universal Exhibtion of 1855 31

Vanadium 102, 154–155
Verne, Jules 3, 8, 13–14, 16–17, 20, 24, 29–30, 38–50, 52–53, 84–85, 90, 98, 117, 121–122, 195, 223, 232–233
Victorian Age 1–2, 7, 9, 15, 24, 47, 50, 53
Vienna Exhibition of 1873 170
Villeroi, M. Brutus 44–45
Volta, Alessandro 130–132, 158
Von Reuter, Julius 140–141

Wallace, Alfred 2
Ward, Captain E. B. 67–68
Washington Monument 123–124
Watanabe, Kaichi 114
Watt, James 180
Weber, Wilhelm 137
Wells, H. G. 11, 17, 23, 30, 41–49, 129, 233
Westinghouse, George 208–222
Wheatstone, Charles 137–138
Whipple, Squire 112
White, Manuel 178
Whitney, Eli 5, 147, 164–165, 179
Whitworth, Joseph 174–175, 179
Wiard cannons 96
Woolaston, William 176–177
Wrought iron 7–8, 28, 33, 44, 54–55, 60, 75, 90, 101, 104–105, 116

X club 11

Yale & Towne Manufacturing 177

Zeppelin 99, 123
Zinc 83, 131–136, 142